TEST AND MEASUREMENTS
PROJECT SUCCESS
WHAT LIES BEYOND LabVIEW
AND TestStand SKILLS

Filipe Altoe

ISBN-13: 978-1497455177
ISBN-10: 1497455170

To the three women who bring real meaning to my life: my amazing wife, Riva, and my phenomenal daughters, Yasmin and Juliet.

It has become appallingly clear that our technology has surpassed our humanity.
—Albert Einstein

Contents

Preface

Along the course of over seventeen years involved in test and measurements project execution at various capacities, this author invested a considerable amount of energy in obtaining the root causes for the failing test and measurements projects as a way of learning how to improve upon the results of future deliverables.

When compiling the results from the numerous root-cause analyses performed for these failed projects, as well as considering what seems to have become the project management industry consensus around why technical projects fail, there is a strong bias toward the so-called lack of well-established requirements as the number one driver for why technical projects fail. Another cause that has statistical relevance is what will be called here poor project planning.

One may say that there is nothing novice and worthy of an entire book to be written based on this conclusion as one probably either came to the same conclusion based on experience or has made use of the vast project management (PM) literature available today in order to arrive at this realization in a very matter-of-fact way. This last sentence is exactly what motivated deeper research that culminated in the work you have in your hands now. The vast majority of the project managers and other members of project teams accept the lack of well-established requirements as their number one reason for why their projects failed, in a very matter-of-fact manner. The majority of the professionals seem to be complacent to the fact that their projects failed due to lack of well-defined requirements in much the same way a cancer patient accepts her fate in face of the incurable disease. It is not uncommon to see project teams placing the blame on the end user for lack of well-written requirements.

They seem almost content when lack of requirements definitions could be assigned as the main root cause for their project failure, as it would be an accepted excuse by the community to a failed project. This seems to be a testament to their competency as there wasn't really much they could do to salvage their projects, as the requirements for implementation weren't well defined to begin with. Their great project management or other project-related skills were put to good use, but there just wasn't anything they could have done better, since, as stated by the literature and accepted by the technical project management community, if the requirements are not well defined, the project will most likely fail.

Thinking in more general terms, this would be similar to saying that the management of technical projects is determined by a component that is outside of the project manager's control, practically being a chance event. If the project team was lucky enough to work with end users who knew what they wanted at the beginning of the endeavor, then the project would succeed; otherwise, fate would take its course.

In the multiyear process of collecting data in search for a potential root cause for why technical projects fail, at project closeout, which is the time to probe deeper into the reasons why the project failed, the almost heretical question of why the requirements were poorly defined would be asked. After receiving a patronizing look from the person who just knows that is the way things go on test and measurements projects sometimes, the vast majority of the time the blame was always pushed to the end user. The end user just couldn't make up her mind in the functionality set and would keep remembering other functions along the course of project execution that would be showstoppers if not included in the initial release.

The real fact of the matter, though, is that regardless of the main root cause identified by the statistics of failed test and measurements projects and who is to blame for it, these projects were, and still are, failing. Millions and millions of dollars are still being wasted in product development initiatives that never see the market light of the day. Excellent ideas never come to implementation fruition due to failed project executions. Hundreds of service-based organizations still go out of business due to their inability to make profits from their project-based service offerings.

There is still an overall main problem to be solved in the industry, which is to change the current statistics of test and measurements (T&M) projects' outcomes in favor of successful execution.

The majority of technical projects and T&M projects nowadays involve a higher number of different technical disciplines than their counterparts of the past. These projects now require a much deeper level of technical skills by project team members in each one of those disciplines than before. Added to that, price pressure has reached an unprecedented high. Moreover, test and measurements projects are usually at the epicenter of new product introduction initiatives. As such, the number of stakeholders that are influenced or touched by a test and measurements project is much larger than it used to be. To make matters more difficult, these stakeholders have a multitude of different backgrounds and professional skills that are not necessarily the same as the project manager of a test and measurements project.

These facts serve as motivators for the project management community to focus growing attention to this ongoing problem. If this trend is maintained without a sound response from the test and measurements industry, the current statistics will tend even more to the side of project failure in the upcoming years for this class of projects.

This can be extrapolated as to create a direct impact in the maintenance of the lifestyle that we have rapidly become used to. Product development organizations will require ever-growing product development budgets, which at some point will have to be rolled into the final end user cost of the off-the-shelf products. System integration organizations will have an even harder time remaining in business as they will either price themselves out of service opportunities due to the added risk costs that will have to be included in the final submitted proposal for technical project execution, or they will corrode their margins to the point of not being viable. It will become harder and harder to materialize into our daily lives the by-products of the latest technological advancements and discoveries. The point of the matter is that this is indeed a serious issue that deserves continuous research and effort.

If the main problems to be solved seem to be lack of well-defined requirements and poor project planning, what can be done to solve for those problems in test and measurements projects? Why don't the existing and, by now, mature frameworks of general project management and systems engineering work for T&M projects? Is it a matter of training of project managers? Is it just a matter of making sure the proper frameworks are being followed, or do we need something different than what is currently adopted by the community? Why doesn't outsourcing of T&M projects to system integrators, the experts in the field of building T&M systems, seem to fix this industry issue?

Answering these questions is what motivated the creation of the TMPM framework, a project execution framework that is tailored for test and measurements projects and that does indeed increase the odds of project success. The work you have in your hands details the process that was utilized for the root causes to be identified, which actually uncovered the real underlying issues that were driving those root causes. Once those issues were brought to life, the TMPM process was derived, focusing on addressing the real T&M project problems.

Introduction

Technological advancements introduced along the course of the last few years have changed the overall expectations of the end user for what can be considered to be a marketable product. As an example, nowadays, end users don't expect a cell phone to just allow them to wirelessly communicate with family and friends, but also to stream a huge amount of real-time data, function as a high-definition TV, and surf the web, all as a basic functionality set. And by the way, they expect to pay just a few hundred dollars for their unit. Price pressure is reaching an unprecedented level on a society that grew accustomed to not having to pay premiums for added functionality. This concept can be expanded beyond cell phones onto a multitude of other examples of the so-called daily life products of our modern society.

As a direct consequence, designing and producing a vast majority of products nowadays involve a higher number of different technical disciplines than their counterparts of the past. They now require a much deeper level of technical skills by product development team members on each one of those disciplines than before. Looking at this scenario from a test and measurements perspective, the complexity of test and measurements systems to test these products has been increasing at a much higher pace than the complexity of the products themselves. And not only is the number of technical disciplines that should now be part of the test engineering umbrella much higher. The aforementioned price pressure from the consumer base forces up the quality levels of the end user units to prevent costly recalls and loss of market share, and test times must be extremely reduced. Test time directly affects production capacity and the capital investment on these test systems, which need to offer reduced cost of ownership.

These technical and price pressures force organizations to implement more and more complex T&M systems, which nowadays can be easily perceived as being part of what is known by the project management community as complex technical projects.

Studies performed by the community have highlighted a very alarming statistic: over two-thirds of all technical projects fail. In fact, this statistic encompasses technical projects of all sizes and industries. The experience gained through roughly twenty years involved with test and measurement systems of increased complexity indicates that this statistic is starting to become valid in the T&M industry also.

Ultimately, the professional project management discipline hasn't been totally adopted by the test and measurement industry, which still is in its infancy when compared with project management performed by technical projects executed in other areas such as construction and IT. Historically, experienced test engineers have been organically made project managers of complex T&M projects without the needed level of preparation in order for them to be successful project managers.

All the factors presented, when mixed together during the execution of complex T&M projects, invariably lead to the same outcome: test department schedules and budgets that are not met, missing test functionality requirements that lead to escapes of bad end user products to the market, inefficiencies in execution leading to raising fixed departmental costs, overall dissatisfaction of organizations toward their test departments...and the list goes on.

If you are reading this book, it means you are somehow involved in the statistics presented above. For the purposes of the discussion presented in this book, a project is considered to have failed if either one or more of the following characteristics can be identified at project closeout: the project went over budget, the project failed to meet its planned schedule, or the project failed to deliver all stakeholders' expected results and/or quality standards, that is, the project failed to maximize the business value to the organization.

The project management community has developed and matured a body of knowledge for managing and executing projects along the course of the last ten years or so. Considerable research and advancements were made around two main frameworks that can be utilized in technical projects today: project management (PM) and systems engineering (SE). A third framework has been gaining momentum in the last few years, especially in the software development industry: agile methodology.

The Project Management Institute, or PMI, is the organ that took upon itself the mission of improving the knowledge base of project management in general. Significant advancements were made toward what is now called the PMBOK®, or Project Management Body of Knowledge.

The PMBOK® provides, at the time this book is being written, a set of forty-two processes and tools that can be utilized by project managers in the activity of managing projects. PMI certainly recognizes the importance of requirements gathering and project planning. It does offer, through the PMBOK®, several high-level processes and tools for facilitating the activities that support the execution and management of projects in general. This addresses, to some extent, the two main causes for technical project failures as previously stated, lack of well-established requirements and poor planning. However, there are some shortcomings when these processes are applied to T&M projects. The main cause of T&M project failures, the lack of well-established requirements, is not addressed in detail by the PMBOK® as an actual methodology to execute appropriate requirements elicitation. There are indeed some high-level scope definitions and requirements gathering processes that provide a somewhat all encompassing set of tools for all types of projects; however, a more comprehensive set of detailed practices has yet to be compiled around the methodologies for requirement gathering to be applied specifically on T&M projects.

The final conclusion is thus that the current project management framework by itself still leaves the door open for failure of technical projects, a fact that is reinforced by the statistics illustrating technical project results. That drives the conclusion for the root cause of this issue not to be a matter of lack of training of project managers alone on best PM practices to be applied to technical projects, but potentially something that needs to be changed and/or added to the PMBOK® framework itself, in order to provide a more specific set of processes and tools targeted to technical T&M projects.

One common action taken by the majority of the organizations in the attempts to mitigate this shortcoming is the promotion of a skilled technical professional into the project management ranks. The rationale is that if a technical resource who understands the idiosyncrasies of T&M projects could be trained to become a project manager, then the problem would be solved.

There are a few problems with that approach. First and foremost, there are some personality traits that are requirements for an individual who is looking to delve into project management, in order for her to be successful at it. In very much the same way a person who is not good under pressure cannot be made an explosives specialist for the military, only individuals who possesses a set of specific personality traits can become a successful project manager. Personalities are formed in the early years of the individual's life, and they are definitely not something that can be easily changed through training during adult life. What that means is that the resource pool is now constrained not only by the level of technical ability the individual needs to posses, but also by specific personality traits that would qualify that individual to be a project manager candidate.

Experience has been showing that it is very difficult finding professionals who posses deep technical skills and also have the personality traits present in successful project managers. Generally speaking, the types of personalities that pursue advanced technical training usually have incompatible characteristics with the ones that would qualify her as a project manager candidate. In summary, finding a skilled technical resource with the personality of becoming a successful project manager is not a small feat.

That being the case, even if an organization is fortunate enough to find a couple of individuals that match those criteria and is able to implement this approach, the organization becomes very heavily coupled with its human resources. This model is not easily reproducible, expandable, or sustainable for that matter. A single turnover event by a resource who qualified as a technical project manager under this model can literately set the organization back several years on its progress, and instantaneously turn success into failure. This shows the overall instability of this approach and that therefore it cannot be considered as a valid solution.

The second framework covered under this analysis is the so-called systems engineering (SE) framework. Similarly to what PMI is for project management, INCOSE (International Council of Systems Engineering) develops and disseminates the interdisciplinary principles and practices that enable the realization of systems. INCOSE also has a body of knowledge that presents a set of processes and tools to guide the implementation of technical projects, much like the PMBOK® for general project management. The point in favor of the systems engineering framework, from here on called SE framework, is that it is dedicated to the execution of technical projects. INCOSE does understand the overall challenges involved in the execution of technical systems composed of custom software and hardware, and it provides best practices as guidance for their successful implementation.

Though the SE framework has the advantage over the PMBOK® framework of being dedicated to technical project execution, providing a focused set of processes and tools for that purpose, there is a fundamental issue with its approach. This framework is heavily based on the so-called waterfall development model.

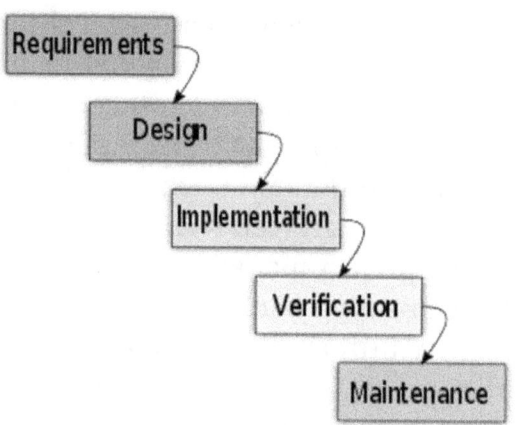

Figure 1. Waterfall Development Model

It is probably worthwhile including a brief high-level description of the basic idea of the waterfall development model in support of the argument that is about to be made. The whole concept of waterfall development is in the idea that the overall project can be broken down into different phases and that work needs to be brought to completion in one phase in order for work on the next phase to be initiated. For instance, all requirements need to be 100 percent defined and set in stone, or, using SE's terminology, baselined, before system design can start. The main argument is that design done on a system with fluid requirements leads to wasted effort and potential rework. There are several instances in the literature that compare the cost of fixing a defect in early stages of the project life cycle versus the cost of fixing the same defect at final stages, the latter being orders of magnitude higher than the former.

Waterfall development thus emphasizes more time spent up front in the project life cycle, highlighting the importance of requirements and design documentation, which certainly carries merit. Other benefits can be perceived when a project is managed under this model, such as:

- It is easier to ramp up new team members in the event of team turnover due to the extensive level of documentation.
- It is simpler to manage since the milestones often include very tangible work products that culminate in the end of a phase.
- It provides very clear and marketable milestones.

In theory, this would be a great answer for the number one reason of why complex technical projects fail; however, in practice, the story is much different than that. For the readers who are familiar with T&M projects and product development efforts, when was the last time the initial set of requirements never changed after the design phase started?

There are several reasons why it is impossible to predict and dictate, with 100 percent certainty, the nature of the entire requirements set of a T&M project. Uncertainties range all the way from changing market conditions, through unforeseen technological roadblocks, and even the human nature of our minds that prevents one to think through every possible scenario and exception case that can be applicable to a given complex system. This leads to the popular phrase "analysis paralysis," whereby an excessive amount of time is spent up front in the requirements gathering phase that ends up generating missed opportunity windows for the implementation of the T&M system.

This is especially applicable to highly regulated industries, such as medical device, pharmaceutical, and government-contracted projects. The fear that stems from project managers of the regulatory bodies responsible for approval of projects within these domains sometimes forces a final product to be released many years after its requirements started being elicited, causing premature obsolescence of the solution in the world of today's fast-paced technology advancements.

Another obvious example of a missed opportunity is in the scenario where the test and measurements system needs to support an organization's NPI, or new product introduction. During an NPI phase, in order for the organization to shorten the new product time to market, the test and measurements system needs to be started well before the device under test, the product to be launched, is fully developed. This causes a typical catch-22 if a typical waterfall methodology is driving the execution of the T&M system.

Under a rigorous waterfall methodology, even when needed requirements changes are identified prior to the final product release, if they come about after requirements baseline, usually the project manager brute-forces the project through completion based off of the obsolete requirements set in order to comply with the waterfall paradigm.

The immediate consequence of this is that the organization may not maximize the full business value that was intended when the T&M project was funded. In this situation, the "success" of the project takes precedence over the success of the organization. As the reader probably recalls, the very definition of project failure presented by this book includes the scenario where the organization fails to maximize the business value of the project.

This also indicates that the SE framework by itself doesn't provide the answers to the proposed questions, and neither does the lack of training and/or monitoring of the SE framework implementation during the execution of technical projects.

The "analysis paralysis" feature of the waterfall development mode drove the creation of what is now called the agile methodology. Agile has gained a good level of popularity among the software development project management community, and it is now making some strides into the overall technical project arena, where the project encompasses multiple technical disciplines and not only software engineering.

The basic ideas of the agile methodology are such that verbal communication is preferred over written documents and working products are the main measure of progress. Also, agile breaks tasks into small increments called sprints. At the end of every sprint, it is expected that the project team has a functioning working product that will get fed back onto the requirements set and will aid the product owner to make more informed decisions about how the system requirements will be prioritized. The learning that comes about with the sprint exercise is used in the tweaking of the requirements set if necessary. In theory, this seems like a silver bullet for the main problem of waterfall development, as one can make the case that it is much more efficient to document something that is already completed. Also, one can argue that it becomes much easier to adapt to the inherent characteristics of complex technical projects, which is the changing nature of its requirements set, as more is learned about the system. With this model, potentially, less effort can be spent on a collection of short-term tasks than on sometimes multiyear-long ones until a workable product is seen. An overall more adaptable process unfolds based on the constant feedback from the many working products that are generated throughout the overall project life cycle.

Unfortunately, theory and practice don't usually go hand in hand in the case of agile development. The lack of documentation in favor of working products more often than not ends up creating too casual of an approach to project execution by the team members, who find in this feature of agile the perfect excuse to avoid less glamorous tasks such as documenting one's work. This creates tribal knowledge and consequently a heavy dependency on individual team members who possess knowledge snippets of the overall project.

Another issue that is somewhat obvious to see is the difficulty of determining an overall project budget and schedule up front. Due to the dynamic nature of the project requirements set in an agile environment, sometimes just a very high-level wish list is created as opposed to an initial requirements set, which makes developing an accurate forecast of schedule, budget, and resources a close to impossible task. This information are fundamental inputs to the organization in its efforts of manage its projects portfolio, as well as setting up an NPI budget. Imagine an executive having the task of prioritizing product A versus B without having access to an analytical set of financial data concerning those projects. Not only that, but since under this model the requirements are so fluid at the beginning, usually system integrators can't really submit a proposal for the execution of a T&M system if this framework is in place.

The last sentence above mentioned the system integrators, and they will be another focus of attention of this book, chiefly around their utilization by clients for the execution of T&M projects. System integrators are herein being defined as professional organizations whose business it is to execute the requirements elicitation, project management, design, implementation, verification, and deployment of T&M systems. As mentioned in the paragraphs above, T&M systems complexity has been increasing exponentially, and as such, it sometimes become difficult for organizations whose main business is not to implement T&M systems to have the needed level of expertise on staff to execute such projects in the time frame its business demands. These organizations usually hire system integrators as the experts in the T&M field in order to shorten the implementation time for the needed systems as well as to reduce the overall deployment risks. Opportunity cost is the name of the game, and as such, hiring an organization whose main business is the execution of T&M systems to run the project sounds like the most reasonable choice.

Contrary to what many may think, this also brings about challenges to the successful execution of the project. Regardless of the level of expertise the hired system integrator brings to the table, as this book will show, there is a set of problems that need to be addressed to foster success of this relationship. The fact that the organization has an expert system integrator in charge of the project is not, in itself, an immediate guarantee of success; however, it is a good first step for sure.

There are five scenarios very common in the relationship between clients and system integrator companies that are important drivers for project failure. Each one of these scenarios will be described in detail in a later chapter.

The fact of the matter is that it is ultimately the client's responsibility to make sure the business value for the T&M system being outsourced to an integrator is realized. Even though the integrator may share some of the same interests, at the end of the day, the integrator is a for-profit business, and its behavior will be highly influenced by what maximizes its success. What this book proposes is a methodology whereby the client places itself in the driver's seat and makes sure there is a clear alignment between the integrator's best interest and the maximum business value for the T&M system being delivered. This also adds substantial value to integrators, as a successful engagement is the number one driver of repetitive business from a client. It is also in the best interest of the integrator that its client maximizes the business value for the system it is contracted to deliver.

Since the two frameworks and the agile methodology have positive processes, tools, and methods, those being somewhat complementary when applied to T&M projects, this provided a good indication that the answer we were looking for would be in some form of a hybrid framework, composed of an adaptation and combination of the relevant processes and tools of PMBOK® and SE frameworks and the agile methodology, into one that is targeted to T&M projects. Moreover, since the involvement of a system integrator is not a guarantee of success for the project either, this hybrid framework needs to address the issues that are usually present when the organization trusts a system integrator with the T&M project implementation.

A more holistic approach to T&M projects is needed, one that would apply not a single person functioning at a capacity of project manager and systems engineer, but a multirole management team. On this new framework, one role would function as the project manager for the project execution itself, bringing all the offered benefits of the PMBOK® framework. Another role would function as the systems engineer, responsible for the technical leadership of the project and bringing with her all the offered benefits of the SE framework. A third role would function as the person making sure the organization business value is being maximized by providing the project manager and systems engineer with all the needed direction, information, and tools in order for them to be successful at their roles, a business analyst of sorts, but focused on T&M. This third role, as it will be seen along the course of this book, will make sure that project success is not just the completion of the initial project scope on schedule and on budget or to make sure the contractual obligations of a system integrator have been fulfilled, if an integration is being used to implement the T&M system. This person will make sure that once the project is completed, it maximizes its business value to the organization. Furthermore, this hybrid methodology should maximize the benefits of the PMBOK® framework, SE framework, and agile methodology, and try to minimize and/or eliminate their shortcomings.

This book has three main high-level goals:

1) To explore the root causes for T&M project failure and to determine the real reasons why these project fail
2) To explore the root causes for failure in engaging a system integrator company
3) To provide a modified framework that facilitates both the successful management of T&M projects as well as the engagement of system integrator companies

This book will present a high-level analysis of the PM and SE frameworks and agile methodology, focusing on their respective gaps and strengths when applied to T&M projects. The same analysis will be performed to cross-reference the aforementioned frameworks and agile with the known issues in engaging system integrators.

This discussion will motivate the introduction of the TMPM framework, which will be presented in detail. The

TMPM proposes a new organizational structure along with a collection of processes, tools, and best practices that address the gaps identified in the conventional PMBOK® and SE frameworks as well as the agile methodology for the successful execution of T&M projects. The TMPM also includes an organizational structure model and life cycle implementation that facilitates the engagement of system integrators for the execution of T&M projects.

This book is targeted to test engineers; professional Lab-VIEW and National Instruments consultants; project managers of test and measurements projects; test managers and any other functional managers that are involved in T&M project execution; engineering and product development executives of service, technology, and product development organizations; and any organization that is faced with the challenging business of implementing and managing T&M projects.

The book is broken down into two main sections. The first section describes the problems that surround T&M project implementation in detail. The second section proposes a new hybrid framework, named TMPM, to address the problem statement specified in section 1.

Chapter 1 presents a root-cause analysis for the two main drivers of project failure, a detailed account for the reasons why T&M projects fail. This analysis is basically the application of a typical problem-solving technique, the goal being to break a large, difficult-to-grasp high-level problem down into smaller, easier-to-answer questions. With the problem at hand, the first step was to perform a thorough data-driven root-cause analysis of the problem in order to find underlying issues.

Chapter 2 presents a detailed analysis of the main issues that drive T&M project failure when system integrators are being utilized to execute projects for an organization. The aforementioned five main issues with this engagement are detailed and complemented with illustrative use cases.

Chapter 3 focuses on a detailed presentation of the current frameworks utilized to manage T&M projects, the ones that are typically used by organizations in the execution of these projects. It also presents a gap analysis of these frameworks, highlighting their strengths and weaknesses when specifically applied to T&M projects.

Chapter 4 repeats the analysis of the current frameworks, but with the focus of cross-referencing the challenges identified in the engagement of system integrators. A similar gap analysis to the one presented in chapter 3 will be determined, but now in relation to the system integrator's engagement challenges.

Chapter 5 summarizes the problem statement presented in section 1 of the book. This problem statement is used as a foundation for section 2.

The second section focuses on presenting the TMPM framework. Chapter 6 introduces the framework. Through a project example, it presents the typical T&M project team structure and the various stakeholders that are usually involved in such projects.

Chapter 7 presents the organizational model that best fits the proposed framework and a description for the multiple roles and interactions between these roles. It also includes a suggested role to make sure the engagement of a system integration company is best aligned with the organization's business value for the T&M system. This chapter will focus on the people element of the typical people, process, and tools combo that defines a given framework.

Chapter 8 focuses on the tools element of the framework. It provides a summarized introduction to UML, Universal Markup Language, and how the UML available diagrams can be utilized to address the root issues for T&M project failure via system modeling. It also suggests a methodology for better manage the T&M project stakeholders.

Chapter 9 brings it all together by demonstrating how the organizational structure and tools presented in the two sections can be put together in a process that is tailored to T&M projects.

Chapter 10 presents an overall summary for the entire book. It ties the problem statement derived in section 1 to the TMPM presented in section 2.

The Test and Measurements Problem Statement

CHAPTER 1:
Why Test and Measurements Projects Fail

An alarming statistical result was presented in the introduction of this book, stating that about two-thirds of all technical projects fail. Two of the main root causes for this statistic were also presented: lack of well-defined requirements and poor planning.

This chapter explores these concepts in more detail and attempts to go deeper into the underlying issues that bubble up as the two mentioned root causes. The goal for this chapter is thus to present a root-cause analysis for the two main drivers of T&M project failure. This analysis will present a set of more fundamental issues that, if addressed, can potentially change this overall picture in favor of successful project results.

Poor Planning

The poor planning root cause will kick off the analysis presented in this chapter. As will be explained in more detail in a later chapter, the main goal of project planning is the generation of a master document, called a project plan, which can be understood mainly as the blueprint of the project. It contains information such as how long the project is expected to last, the resources it will utilize, how much it will cost, the project main risks, and the high-level activities that need to be executed by the project team in order for the project to fulfill all its determined objectives.

That last phrase is extremely important, as it partially states the very definition of project success: the fulfillment of the project objectives. This is very important to highlight since, amazingly enough, a good number of T&M projects that are started don't have a good set of project objectives. It doesn't matter the angle one utilizes for analyzing project success; if the project doesn't have an *a priori* definition of what success means, the exercise of determining if the project was successful or not becomes a battle of opinions as opposed to a data-driven analytical process.

What this concept suggests is that every project must carry an agreed-upon set of high-level objectives to be met, in a way that project success can be objectively measured at its closeout phase.

What this indicates is that the criteria for overall project success have a common set of components, an agreed-upon and planned schedule and budget, and the high-level business value it will realize, as well as a project-specific component, the project objectives. Cross-referencing this again with the overall definition of project success, it is easy to see that project objectives need to include all stakeholders' expectations and acceptable quality standards.

In reality, only a few project managers actually make sure projects have a valid set of project objectives before they are started. They usually get caught by the pressure of making progress on project execution and jump to the project planning activities, working in determining the project's work breakdown structure and the follow-on tasks that lead to a project plan. The main problem with this approach is that the planning activities, regardless of how thorough they are carried out by the project manager, are started without the right focus given by the set of project objectives. This opens up the project objectives list to the interpretation of both the project manager and the project team, while attempting to execute the planning activities, and later on the project execution activities.

The project manager's main job is, again, to make sure the project meets its objectives, not defining its objectives along the course of the project. Regardless of the project manager's opinion about what should constitute the project's objectives, their definition is beyond her role. A good project manager identifies gaps in project objectives, brings them to the attention of the project sponsor, and makes sure they are all addressed before officially starting the project planning.

If the project manager fails to do that, at that point, the project is already headed to disaster. This leads us to our first identified underlying root cause of a poor planned project: *Lack of Established Project Objectives.*

As it mentioned above, it is not the project manager's role to determine the project's objectives, but to drive the project to meet them. The multiple project stakeholders are the ones who need to provide input to fill out the project objectives. This can be identified as the first issue with the PM framework. There usually isn't a role in T&M projects specifically defined to make sure there is a level of facilitation to gathering and executing the final compilation of the information that will define the project objectives.

One may argue that this is the project manager's job. However, in reality, since a T&M system usually executes a pivotally important function for the overall organization, the definition of the project objectives needs to be done by a person who can more freely float around all organization departments. This will allow this person be appropriately identify all stakeholders, another issue that will be seen below, and make sure the stakeholders' data are properly compiled as the set of projects objectives.

The T&M PM, whether a member of the organization's staff or a system integrator, usually doesn't have that flexibility or know-how.

One very typical problem that usually surfaces in late stages of the project execution is the identification of missed or misinterpreted objectives and the consequent misalignment between the project plan and the real objectives. As mentioned above, there should be activities of stakeholder identification and extraction of the information; however, since complex technical projects usually involve objectives from multiple disciplines, such as sales, marketing, organization strategy, R&D, and others, the stakeholder pool is composed of individuals with multiple backgrounds. On top of that, individuals who fill those roles in organizations that will execute technical projects do not necessarily come from technical backgrounds. This leads us to an underlying issue that drives the lack of established project objectives: *Poor Communication with Stakeholders.*

This will force a sometimes technical project manager to communicate with individuals from other organizational roles that might not be technical during the gathering of the project objectives. Most of the time, there is a gap that needs to be bridged when that is the situation. The bridging of this gap is the subject of a later chapter.

Miscommunications with stakeholders will lead to misinterpretation of gathered project objectives, which will lead the project manager back to the trap of interpreting the project objectives, and consequent failure to meeting the real objectives that were not properly captured.

Have you ever been in a project where you believe the project delivery will be a slum dunk only to find out that a showstopper function was missed altogether, which prevents the customer from taking delivery? Assuming in this situation that the project execution was properly managed to the existing requirements and traceable back to the gathered project objectives, this will most likely have a single root cause: an important stakeholder was not identified in the early stages of the project and therefore his expectations of the project were not properly captured as an objective. Since requirements are derived from higher-level objectives, the true underlying issue might not be lack of established requirements, but rather a lack of established and validated objectives.

This takes us to another underlying issue around lack of established project objectives: *Failure to Identify All Stakeholders.* This is another capital mistake that happens more often than it should and also drives the project toward a cliff, in the very early stages, and usually is only identified when it is already too late for recovery. The result is project failure due to poor planning, regardless of how well the project plan was carried out.

Besides the fact that the project manager may miss some objectives from stakeholders for the two reasons that were presented above, she also needs to make sure that both high-level business and technical project objectives are identified and captured. One characteristic of technical professionals who become project managers is that they will pay extra attention in making sure the technical objectives are captured and spelled out; however, they are usually weak in the facilitation of gathering the business objectives.

The business and technical project objectives can be referred to as the high-level project scope of work. Therefore, we can list the next underlying issue that can drive poor project planning as *Lack of High-Level Scope of Work.*

Please note that I refer here to a high-level scope of work and not as the final scope of work that will be one of the outputs of the project planning activity, which are determined by progressive elaboration of the high-level scope of work plus requirements gathered. A high-level scope of work, though, is fundamental to guide the project manager during the intermediate project planning activities and should also serve as input to the requirements gathering process.

One final underlying issue that drives incomplete project objectives is related to quality standards, which is part of our proposed definition of overall project success. It is of paramount importance that the project objectives also carry a very comprehensive description of what defines acceptance of the high-level project artifacts that will fulfill those objectives, qualifying them as a successfully implemented at project delivery. This means that it is not only important to define what will be the high-level scope of work from all relevant stakeholders and make sure communication with these stakeholders is not creating any misunderstanding around that scope, but also to define what will be the acceptable standards for the deliverables that implement that scope of work. This brings us to the next underlying driving issue for lack of established project objectives: *Lack of Acceptance Criteria.*

Have you ever been in a project where there was conflict between what the project team members considered to be a complete deliverable and what the receiving end person of the project considered to be her standard for an acceptable deliverable? That most likely happened due to either some ambiguity or total lack of a comprehensive acceptance criteria for that given deliverable. The best way to end conflict is to refer to some sort of signed contract language that is put in place in the very early stages of the project life cycle.

Once the quality standards are defined and agreed upon in advance, the planning activities will have to take those standards into consideration with the project plan. Imagine for instance a project in which quality standards must be compliant with the strict FDA (Food and Drug Administration) regulations and a second project where no regulatory agency will be involved. The project plan for the first project will have to include a much more comprehensive verification and validation plan than the one for the second project, affecting its overall schedule, cost, resources, risk, and other parameters that are considered during the project planning activity. Now imagine that the acceptance criteria for the first project fail to capture that the project needs to withstand a FDA standard quality audit. This is a really gross thing to miss, but it helps to reinforce the idea.

Should this distinction not be known prior to the project planning activity kickoff, the project plan will be created based on either an assumed quality standard by the project manager and/or the project team, or one that is probably ambiguous. This will most likely lead to an eventual rebaselining of the project at some point during its execution, leading to project failure.

Since project planning is a task that is carried out in the very early stages of the project life cycle, stakeholders, project manager, and project team still don't have a chance to learn much about the project in general. Usually, as the project is carried out, the overall knowledge about it increases, leading to a situation where more accurate forecasts and decisions can be made. However, the project needs to start somewhere. At very early stages, accurate predictions and learned facts that come with the effort put into project execution are traded off by assumptions. One of the biggest mistakes I see happening on T&M projects is that the assumptions made at project initiation are made by the project team and are not validated with the project stakeholders.

The project team more often than not will make assumptions not only about the project objectives, as mentioned during the discussion about the underlying issues for lack of established project objectives, but also about other areas of the project. This is not a completely forbidden act for the project team; however, any and every assumption made needs to be properly validated with the project stakeholders. This leads us to the next underlying issue of poor planning: *Lack of Validated Assumptions.*

Assumptions that are not validated at the early project initiation phase may be the kiss of death for technical projects. It will lead the project manager down a fictitious path for the project planning, which will provide an output that is based on assumptions that are not necessarily valid from all stakeholders' perspective. Again utilizing the definition of project success, which is that the project must accomplish all stakeholders' objectives and quality standards, the utilized assumptions need to be validated with the stakeholders; otherwise the project is at risk of accomplishing the project team's objectives and not the stakeholders'.

This is a very subtle issue and sometimes hard to identify in a project until it is too late. The project manager for technical projects should always be in a state of alert for assumptions being made and make sure stakeholders are in agreement with them.

The following underlying issue that will be described is one of the strongest drivers of project failure via poor planning and should therefore be given the appropriate level of attention: *Poor Risk Identification.*

Overall risks can be classified in two main areas: unforeseen risks and known risks. Common sense suggests that the higher the ratio of known risks over unforeseen risks the smoother the project execution becomes, and that is indeed the case. Unforeseen risks can totally derail the project beyond the point of recovery, depending on its overall impact. Now, think of a complex T&M project you have been involved in where a thorough risk identification process wasn't properly implemented. This project most likely had a very low known over unforeseen risk ratio, leaving the project very vulnerable to what is usually called chance events, or plain bad luck.

What I am proposing here is that luck in T&M project execution doesn't really exist, just bad risk identification and planning. Statistically speaking, if the risk identification process is given the proper attention and, at the minimum, the medium- and high-probability events are identified and proper risk response plans are put in place to handle potential realization of those risks, odds are that this project will most likely not have to face any of the so-called chance events. Generally speaking, under this scenario, totally unforeseen risk realization has very low probability. Even if a low-probability event of unforeseen risk comes to realization, at that stage, the practice of allocating a management reserve to the project is usually enough to cover these low-probability unforeseen risks, and the project still ends up being a success.

Based on the proposed arguments, risk identification, much like the lack the issue of lack of established project objectives explored previously, can be broken down as a set of underlying issues that drive its occurrence. On a complex T&M project, one can classify its risks of belonging to two main categories: business risks and technical risks. It is important for the project manager to recognize this breakdown and to give proper attention to both types of risks, involving the appropriate stakeholder on the activity of risk identification for both.

One of the common mistakes made by technical project managers is that risk identification is mainly trying to determine the technical risks of the project, and very little attention is paid to the business risks. As it was described earlier, under that scenario, those business risks become part of the unforeseen risks that will have high probability of derailing the project. How many times did you hear a technical project manager state, "Who could predict the project was going to be terminated by the executive team due to lack of sales?"

Well, in this case, if the appropriate stakeholders were properly involved as part of the risk identification activities, the project manager could probably have identified this as a medium- or high-probability business risk. This in turn could have been properly captured as a legitimate project risk, which would have provided extra information to the project sponsor, who at that point might have decided to either postpone or kill the project altogether, saving the organization thousands or even millions of dollars, depending on the nature of the project. On this instance, the project manager would probably try to defend herself by saying the project failure had nothing to do with her skills as a project manager, but in this case my opinion is that she did indeed fail as a project manager because she didn't do proper risk identification.

This example takes us to the first underlying issue that drives poor risk identification: *Unbalanced Business and Technical Risk Identification.*

A well-balanced risk identification exercise always considers both types of risks, business and technical, for complex technical projects. A balanced risk identification activity needs to apply a comparable amount of energy on the identification and planning of both types of risks. This suggests that only a well-rounded business and technical project manager can potentially execute such well-balanced risk identification. This will be explored in a later chapter; however, this last statement, when taken together with the previously stated fact that it is an incredibly difficult task finding a professional who has, at the same time, the personality traits of a successful project manager as well as deep technical skills, provides the first hint about how the organization should shape up for the successful execution of a complex project. This is important enough to have an entire chapter devoted to it, and it will be addressed in due time.

Drilling deeper on potential underlying causes for an unbalanced risk identification exercise, the first cause that comes to mind is *Not All Appropriate Stakeholders Were Involved in Risk Identification.*

No matter how centralized the information is in an organization, a statement that is always valid is the following: no single person has all information about all possible areas of an organization. Only someone involved in sales has all the underlying details that are related to sales; only a marketing professional has all the underlying details related to marketing; and so on. Even in the event of a very hands-on company president or CEO in charge of an organization, it is humanly impossible for that person to know every nuance of all different areas of the organization.

Failing to involve a domain expert on the risk identification process of a complex technical project will most likely lead to risks that could have been identified and planned for and that will now reside in limbo as unforeseen risk events. It is paramount to remember that not because a risk wasn't identified it doesn't exist. This leads to the problems that were already explored previously. Failing to involve relevant stakeholders on risk identification will shield the project team from being exposed to a perspective that could have been fundamental in identifying a high-probability risk that may drive the project to failure.

A corollary of the underlying risk presented above is a situation whereby the project manager feels that she can do it all, without involvement of any other stakeholder. This is close enough to the issue above, but since I have identified it as a recurring item among T&M project managers, I have decided to make it stand on its own as a separate issue that needs to be addressed.

This is most common with technical professionals that are made project managers. They usually feel they are the most qualified person in the project team to identify risks. There is actually a combination of issues with this scenario. First and foremost, the project manager is failing to balance business and technical risks by believing she is the most qualified person to be identifying risks because she is a skilled technical professional. She is falling into the trap of not even identifying that there are business risks associated with any complex T&M project.

The second issue is the fact that even during the identification of technical risks, the power of a team is much higher than the sum of the number of its members. There is a concept called "masterminding," which is not very well understood and/or applied by the technical community.

Napoleon Hill, author of the book *Think and Grow Rich*, first defined masterminding as a "coordination of knowledge and effort, in a spirit of harmony, between two or more people, for the attainment of a definite purpose." Hill's concept of the "Master Mind" was inspired by Andrew Carnegie, that wealthy steel magnate. According to Hill, "Mr. Carnegie's Master Mind group consisted of a staff of approximately fifty men, with whom he surrounded himself, for the definite purpose of manufacturing and marketing steel. He attributed his entire fortune to the power he accumulated through this 'Master Mind.'"

Since the publication of *Think and Grow Rich* in the 1937, the idea of mastermind groups has grown and evolved to become a staple tool of successful individuals and even organizations. One can list the following benefits of a mastermind group:

1) A group of individuals functioning to achieve a single objective.
2) Differing perspectives, input, and feedback.
3) The mastermind team can bring resources and connections to the table one single person might not have had on her own.
4) The team benefits from accountability and inspiration from the whole group, thus enabling you to maintain focus in achieving your goals.

I have seen very difficult, "unsolvable" problems solved by the power of masterminding, and this is something that even experts on their fields can benefit from.

Another common mistake done by a technical project manager is: *Not Making the Risk Register a Living Document.*

Even if the technical project manager sees the value of risk identification, most of the times risk identification is done only once at the very early stages of the project planning and then baselined as if it would be immutable from that point onward. This couldn't be more dangerous. Just think of how much more is learned about the project when its execution progresses versus the information the project team and the stakeholders had at project planning.

It is just silly not to take advantage of the extra technical information set that is obtained during the project execution. This expanded information set can easily change the overall technical risk landscape by validating identifying risks, tweaking probabilities and impacts of other identified risks, invalidating other identified risks, and identifying new risks that couldn't have been foreseen with a smaller information set. Always remember that the name of the game here is to reduce the unforeseen/identified risk ratio; therefore, continuous risk identification, analysis, and planning effort need to be executed throughout the entire project life cycle.

Another point to be made on this topic pertains to the business risks. Business risks are, by nature, volatile as they are usually tied to market conditions. One doesn't need to make a point on how volatile the market conditions are; therefore, by proxy, the business risks are in a constant state of flux.

This indicates that most of the time, risks identified in the past might carry different probabilities and impacts, if they are still valid at all. Furthermore, new risks might have become visible under the modified market conditions; therefore, they need to be captured.

To summarize this topic, one of the most common mistakes done by technical project managers is to utilize project team meetings as a "go around the table" status type of meetings, whereby the team members are asked about the status of their tasks and what they are planning to do next. Status information can be gathered and communicated by taking advantage of lower overhead methods that need to be spelled out in some sort of communications management plan.

Team meetings should be considered a precious opportunity for collaboration and masterminding, not as a mere status-gathering tool. It is usually very expensive for the project to gather all team members in a room for a couple of hours; therefore, maximum benefit should be taken from that activity. As mentioned, risk identification is one of the biggest drivers of poor planning and consequent project failure. Also, risk identification needs to be a continuous activity along the course of the entire project life cycle. Marrying these two concepts, we propose that constant risk identification in a masterminding format needs to be the central point of any team meeting agenda.

The frequency of these meetings will depend heavily on the overall complexity of the T&M project, as well as the overall level of risk identified for the project. It is common sense that it would be unnecessary to gather the project team twice a week for two hours to execute a simple project, and on the other side of the spectrum, to gather the project team for a one-hour meeting once a month for a complex T&M project. The frequency of the meetings is also something that needs to be planned and included as part of the communications plan.

I will direct the next issue to service-based organizations. These organizations, more often than not, need to come up with a fixed-price proposal including some sort of high-level implementation schedule execution of a given project for their client. It is intuitive to understand that this is a very risky business proposition in itself. It is usually very hard to even get a detailed enough understanding about the scope of work in order for an accurate proposal to be created.

One can possibly argue that the level of understanding of the scope of work of a project is directly proportional to the effort invested in the proposal generation. Though that is a true statement, one should also keep in mind that these organizations, being service based, basically make their money by selling their staff's time. Therefore, time spent on tasks that are not generating revenue directly affect the overall organization's bottom line. Furthermore, proposal generation activities often are not paid for by the client, and usually these organizations are part of a competitive bidding procurement scenarios. What this means is that all the work invested in generating the proposal can potentially lead to no revenue whatsoever if the organization is not awarded the contract.

With all this in mind, the question of how much time is the ideal time to be invested in proposal generation is an open problem in the service business for T&M projects. On top of this, not many companies start a formal risk identification process as early as the precontract phase because they see that as a project management function, therefore belonging to the project execution once the contract has been awarded.

What ends up happening is that since a poor risk identification process was executed, the submitted proposal usually fails to take those risks into consideration when determining the project budget and high-level schedule. There are two direct possible outcomes of that fact:

1) Service organization overshoots the customer budget by padding proposal with excessive management reserves, losing the contract opportunity to a competitor

2) The winner's curse: service organization is awarded fixed-price contract, but unforeseen risks are realized and the real project budget is much higher than the price the contract was awarded for. The final outcome is that the organization loses money.

A corollary of the second possible outcome is that the service organization realizes the proposal was underbid and attempts to cut corners on project execution, sacrificing the quality standards, and consequently fails to deliver the project objectives. Worse yet, the service organization might choose to find contractual loopholes in order to ask the client for more funds in the form of engineering change orders, damaging the overall relationship with the client and preventing repetitive business to be made with that same client.

Lack of repetitive business leads to a more expensive sales process, since now the sales force needs not only to work on proposals but to actually prospect new leads of potential new clients to fill the sales hole left by the poor execution and customer relations. This raises the organization's overhead, which in turn makes it harder operationally for the organization to survive.

This discussion leads to one of the issues identified specifically for service-based organizations: *No Formal Risk*

Identification Process Done as Part of Precontract Phase. This issue will be explored in more details on the chapter specific to working with system integrators; however, for now, I will leave the reader with the idea that it is the client's responsibility to make sure appropriate risk identification and scope of work definition is done prior to the contract being put in place. Even in a fixed-price contract scenario, where the integrator takes all the financial risks, it is extremely expensive to the client to have a failed T&M project.

Another issue that is worth mentioning on this section about T&M projects is that for the most part, technical project managers don't really understand the importance and the potential impact of the overall risk identification process as part of the project planning. The main reason for that is failing to understand that the project manager is ultimately responsible for the success or failure of the project. The lack of understanding that the so-called chance events that derailed the project are actually realized risks that were not identified is a tough concept to get through to project managers.

I will list here the last issue that drives poor risk identification: *Lack of Understanding of the Importance of Risk Identification.*

The inherently human reaction of self-preservation prevents project managers from accepting total responsibility for their failed projects. They prefer to sweep their failures under the carpet of the chance events than actually accept that something they could have done better would have turned project failure into success. This is obviously a very difficult thing to change, and it needs to be driven from the top of the organization down to the project managers.

There is one last issue that is worth mentioning in this section that does create a very serious impact on the project planning activities, and consequently the overall project performance: *Lack of Stakeholder Buy-In on the Plan.*

This is also a common mistake done by technical project managers, and it basically consists of not pursuing a thorough review and subsequent buy-in from the project stakeholders. The process of securing buy-in from stakeholders usually forces buried issues to the surface, and it forces stakeholders to put their names in the project plan as agreeing to everything that it states.

Like any other document that requires signatures, a project plan that is signed by every stakeholder who participated on its conception, especially during the risk identification and planning phases, forces the stakeholders to either agree to what is presented or to voice their reasons for disagreement. The later usually indicates that some risks were not properly identified during the risk identification process, or objectives were not properly captured.

This shouldn't be perceived as an exercise so the project manager has a document that can back her up in case of project failure later, but as a true value-added task that will force the appropriate level of due diligence from all project stakeholders. This not only gets participation from the multiple perspectives that will be involved in a project, but it is also as an extra step in order to further bulletproof the project plan and raise the chances of overall project success.

An extra iteration in the project plan that might be generated by the need for official buy-in, represented by stakeholder signatures, might look, at a first glance, as an unnecessary project delay. However, this is a win for the project manager since this exercise will force a more detailed activity focused at whatever issues are the reasons for the disagreement. At the end, a stronger project plan will be the output of this phase, and since poor project planning was identified as one of the root causes for the core problem we are trying to solve in this book, this delay should be welcomed by the project manager and overall organization. This doesn't mean time is being wasted but, actually, saved. Any extra hour spent on planning will save many times over the time spent later in the project life cycle, as objectives have been properly defined and a good identified risk register is obtained; these set the ideal foundation for project execution.

Figure 2 summarizes the breakdown of all identified underlying issues that culminate as poor project planning. This picture also shows the relationship between each identified issue with its upper-level direct effect.

Figure 2. Underlying Issues for Poor Project Planning

The next section will present similar root-cause analysis of the main driver of complex T&M projects' failure: *Lack of Well-Defined Requirements*.

Lack of Well-Defined Requirements

This section will present the root-cause analysis for what has been established by the project management community and by empirical observations as the main driver of complex T&M project failure: *Lack of Well-Defined Requirements*.

Following the same rationale as the analysis made for the other main driver of complex project failure presented in this chapter, the lack of well-defined requirements can be broken down further into four main driving forces. These driving forces will be explored in detail in this section.

The project objectives of complex T&M projects, as was seen previously in the course of this chapter, determine the high-level business and technical goals for the project. It can be loosely seen as the stakeholder's wish list end results for the project deliverables in both business and technical areas. The project requirements can be seen as a deeper level of abstraction of the project objectives. This deeper level would define in finer details what the project deliverables must implement. It thus becomes natural to see the requirements as having the project objectives as their foundation. What this sentence means in practical terms is that much like a building that needs a solid foundation, project requirements need solid project objectives to be used as starting and arrival points.

The above paragraph leads us to the first identified driver of poorly identified requirements: *Lack of Well-Defined Project Objectives.*

Much like the popular saying "garbage in, garbage out," any requirements-gathering activity that starts off with poorly defined project objectives will provide poor requirements. It is of fundamental importance to start any requirements-gathering exercise with the finish line in mind, how these requirements will allow the project team to deliver a work product that meets all captured stakeholders' objectives. This underlying driver also shows how the two main drivers of project failure discussed in this book are intertwined. Even though the main driver of complex project failure might be lack of well-defined requirements, there are some components of bad planning that reflect in the requirements-gathering activity as well.

There are two underlying issues that are common to both poor planning and lack of well-defined requirements, the first being *Failure to Identify All Stakeholders.*

This is also a hard problem during requirements definition, even in the event that the project objectives were successfully acquired by proper identification of all stakeholders. The main issue here is that project objectives and project requirements, though very much connected, live in two different spheres. The project high-level objectives for complex T&M projects, though certainly including some high-level project requirements, mostly belong to the business category.

Conversely, the project requirements sit mostly in the technical domain, although they definitely carry a component that is business related. Please note that when I say "technical domain" in this case, I am not referring to functional requirements alone, but also the nonfunctional requirements, as we will explore later. For this reason, there is a large component of project requirements that are unrelated to the stakeholders that were defined as the appropriate group to participate in the definition of the high-level project objectives. Figure 3 illustrates this idea.

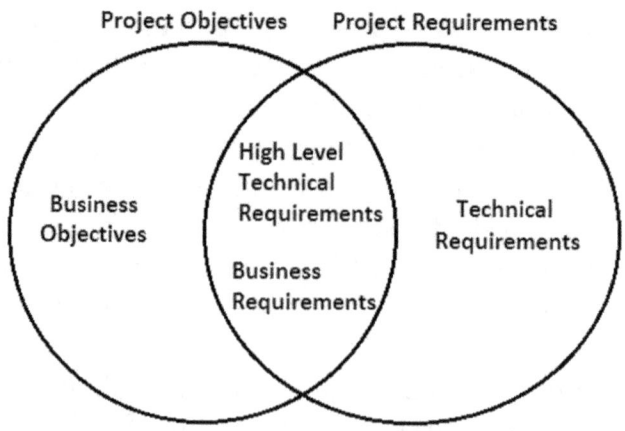

Figure 3. Project Objectives and Requirements Spheres

By analyzing figure 3, it becomes clear that although there is certainly an intersection group of stakeholders that should participate both in the project objectives definition and the lower-level project requirements definition, there are stakeholders for the project objectives sphere that are not related to the requirements-gathering activity and vice versa.

What this means is that a similar analysis to the one performed to identify the stakeholder registry for the project objectives definition during project initiation needs to be executed during the requirements-gathering activities during project planning. We will hold off on the presentation of a suggested process of properly doing this identification until a later chapter. For now, just know that failure to identify all stakeholders during requirements gathering is an issue on complex T&M projects that need to be addressed.

The second driver for a lack of well-defined requirements that is similar to a previously identified driver for poor project planning is the *Communication with Stakeholders.*

As it was presented in previous paragraphs, the requirements-gathering activity is in a different sphere than the project objectives. Also, it was presented that this activity is much more technical than the project objectives definition. As it was mentioned previously also, one of the problems identified as a root cause for communication with stakeholders during the project objective identification process was that, most of the time, technical project managers don't really have advanced business acumen, as they have been promoted from the technical ranks into the project management world.

In the requirements-gathering case, the reverse problem is usually identified. Project managers who did execute proper project objective identification due to their advanced business acumen usually fail to understand deep technical issues and sometimes even the specific technical lingo used by the stakeholders. This leads to a translation error between the information passed back and forth between the project manager and the stakeholders.

The conclusion to be drawn is that proper project objective identification and requirements gathering can be almost mutually exclusive; technical project managers can properly perform the requirements gathering but usually fail in the project objectives activity, and more traditional project managers excel in the project objectives identification but fail in requirements gathering. This is obviously not applicable to the rare instances where technically skilled resources have also business orientation as part of their personality and interests. However, we also explored previously in this book that this case cannot really be considered a solution as it makes the organization to become too heavily tied to its human resources pool.

The next issue to be explored and qualified as a driver of poorly defined requirements in complex T&M projects is what will be called here *Lack of Attention to Nonfunctional Requirements.*

This is most typical in T&M projects being managed by technical project managers. Due to their technical backgrounds, these project managers usually excel in the identification of functional requirements as they are most likely closer to their technical areas of interest. What usually happens in this case is that these projects have well-defined technical requirements for their expected functionality, but fail due to nonfunctional requirements that were either neglected or not properly explored in detail.

As the old saying goes, "The devil is in the details," nonfunctional requirements, often seen as unimportant details by technical project managers, can kill complex T&M projects.

This also indicates a somewhat mutually exclusive relationship between the functional and nonfunctional require-

TEST AND MEASUREMENTS PROJECT SUCCESS

ments when handled by a single person. Technical project managers will most often overlook nonfunctional requirements and excel on technical requirements gathering, and traditional project managers will most often not be well equipped to handle technical functional requirements definition but do a good job of defining the nonfunctional requirements of a technical project. The proposed TMPM framework addresses this issue; it will be presented later in this book.

The next driver for lack of well-defined requirements that will be presented is not necessarily unique to complex T&M projects as much as it is a general problem that affects any complex project. Granted, complex T&M projects usually have a larger requirements set than nontechnical projects since technical specifications of functionality are a component of the overall functional requirements, and that alone usually carries pages and pages of functional requirements. This is especially so with T&M projects that involve a large suite of engineering disciplines as part of the solution, such as software, electrical, mechanical, RF, and others.

For such projects, it is usually useful for the requirements gatherer to break the activity down into multiple requirements-gathering activities for each one of the subsystems, as it will presented as part of the TMPM framework, and also to make sure that the requirements for the integration between the subsystems are also determined as part of the process. What this indicates is that the process of gathering requirements for a complex T&M project can be a very daunting task, and as such, it is not uncommon for some requirements to be missed in the process. This is a chronic problem that is identified here as *Requirements Errors of Omission.*

Errors of omission can be classified as any requirements that were missed during the process of gathering requirements. Intuitively, it is somewhat easy to see how this can easily be the case for complex T&M projects that involve multiple engineering disciplines as illustrated above.

The traditional process of opening a word processor and typing up requirements might be sufficient in the case of smaller and less complex projects, however it usually leads to errors of omission on complex technical projects. Regardless of the technical aptitude and business acumen of the person executing requirements-gathering activities, a more friendly process than the aforementioned word processor needs to be used in order for errors of omission to be minimized.

The TMPM framework includes a comprehensive requirements-modeling process, based on UML language that will be explored in details at a later chapter. In short, this method is a visual representation of the multiple layers that compose a T&M project, allowing the requirements-gathering team, stakeholders, and project manager to visualize all potential exception cases that are usually missed and cause errors of omission as defined above.

The next driver of poorly defined project requirements is probably the one that is heard the most from project managers: *Users Don't Know What They Want.*

This driver is especially interesting; it deserves attention and a deep analysis. At first glance, it is natural to assume this is an issue with the end user and something that the project team, including project manager, cannot really do anything about because it is an external issue. Being an external issue, the project team therefore doesn't have any control over and cannot prevent projects from failing if their project falls within that trap.

This is especially apparent on complex T&M projects whereby the end users usually have only an initial idea of what the final deliverables should look like at the early stages of the project planning. As the design and later the implementation phases start, users start to have a better idea about their desired outcome. Usually, by that time, the project baselines of scope, budget, and schedule have been already set, and the project outcome cannot be anything else other than failure. This is an obvious underlying issue of lack of well-defined requirements in the project planning. However, I propose that this doesn't necessarily need to be treated as an external issue and consequently out of the control of the project team.

Further analysis to this situation was done and underlying situations were determined that surfaced as lack of knowledge on the user's part on what should be the desired outcome for the project deliverables. This analysis will focus on complex T&M projects; however, a similar analysis can be expanded to all projects that suffer from this problem.

Going back to the root of the issue, T&M projects objectives are usually determined by a high-level market-driven void that needs to be filled by the given organization. Market needs can be generally from one of the following categories: technological advancement, legal implications, and regulatory standards.

The user of a T&M project might or might not have a technical background that would allow her to foresee the exact format they would like to see for the final project deliverable. By the same token, the end user might not have a business background allowing her to also have a very detailed view of the final project deliverable. It is one of the project team's main jobs to facilitate the task of having the end user see as much of a final picture of the project deliverables as possible.

Again, utilization of the traditional word processor way of gathering requirements doesn't foster a proper environment to execute such activity. One of the reasons why users seem to make up their minds later in the project life cycle is because at that later stage, they usually have the opportunity to see something in front of them as a semifinished product. At that point, they may realize that what they are seeing won't meet the high-level project objectives, either on a technical or business level. Thus, their minds seem to "change" at that stage in relation to the original requirements.

In fact, what happens is that they never actually had a chance to visualize and create a mental three-dimensional image of the project product until that point. That is usually the first time they were able to mentally connect what they are seeing as the product with the project objective. At that stage, it is indeed a chance event to have a deliverable that meets the project objectives as the end users didn't really fully participate in the process of defining the requirements, even if they were asked their input on determining the requirements.

End users seem to all of a sudden magically know what they want when they are looking at the product, and usually it is not what they are looking at. That is when frustrations reach their peaks on both project team's and end users' sides. That is when the end users are usually blamed for not knowing what they want.

A successful project team provides the appropriate set of tools that allow the requirements to flourish out of all stakeholders. This is when some hybrid framework between agile, PM, and SE would come in extremely handy. What the TMPM framework implements is, as it will be seen in later chapters, an initial requirements modeling based on UML followed by an agile implementation that allows the end user the opportunity to see deliverables early in the project life cycle.

Modeling is a concept that has been evolving in the past decade or so due to the technological advancements around the PC and overall processor power. Computer-aided design, or CAD, tools were one of the precursors of system modeling. They allow three-dimensional mechanical systems to be modeled in the computer before being fabricated. Electrical circuit modeling tools also allow electrical engineers to model their electrical designs prior to PCB fabrication and assembly.

Modeling was taken to the next level by the systems engineering community on what is now known as SysML, which serves as one of the bases for the proposed TMPM framework. It does solve multiple problems around requirements gathering, namely, stakeholder communication, errors of omission, and the fact that the end user seems to never know what she wants. It presents a visual representation of how the product of the project will look like prior to the design and implementation of it.

Since the modeling activity focuses on the requirements; it allows the user to navigate through all potential exception cases during the requirements-gathering process. It also allows the person who is gathering the requirements to minimize errors of omission by being able to explore all potential corners of the requirements-gathering process, with the aid of a visual representation of the system. Finally, it bridges the existing gaps between stakeholders due to potential differences of backgrounds and training.

Another driver of poorly defined requirements worth mentioning in this text is *Lack of Attention to Historical Requirements.* The majority of the time, every new complex T&M project being executed by a given organization is handled by the project management body, not only as a new project but also as if the organization has never executed any project even remotely similar to it.

In reality this is not usually the case. Since the organization's main business is usually what drives the creation of new T&M projects, it is very natural for these projects to be somewhat similar in nature. As such, the person responsible for gathering the requirements of the new project could immensely benefit from starting the requirements-gathering activity with something other than a blank canvas.

Imagine how much easier the task of gathering requirements for a new project would be if the party responsible for the activity had at her disposal all the requirements gathered for a multitude of previous projects executed by the organization. This would include not only the approved requirements that were included as part of the final baselined requirements set, but also the ones that didn't make the cut.

This person would be able to identify existing requirements that are common between her new project and similar ones executed in the past, jump-starting the overall process. Furthermore, it would be of much value for her to analyze the requirements that were not approved on previous similar projects but that would be applicable to the existing project. This activity would immediately alert the requirements gatherer to potential quicksand areas that trapped past projects and that could potentially come back to haunt the new project.

This discussion leads to an underlying issue that drives the lack of attention to historical requirements: *Lack of a Requirements Management System.*

A requirements management system is basically a database that keeps all requirements elicited for past projects and that can and should be used during the life cycle of active projects as well. This system keeps requirements that were approved and as well as requirements that were rejected. This system usually also stores information about why requirements were rejected, as well as general comments that might be relevant information for new projects.

The next underlying issue to lack of attention to historical data that can be listed is *Lack of Organization Process Assets.*

Another way of understanding organization process assets is to call them templates. Templates showing a high-level roadmap for the process of gathering requirements that are applicable to a given organization are always very advantageous for the requirements-gathering person to have.

Requirements templates can show some requirements packages that most likely will contain requirements to be gathered for any project for the given organization, such as regulatory requirements, chemical process requirements, user interface requirements, etc. These packages are obviously very dependent of the type of business the organization is in, and they can serve as a guide for the requirements-gathering person to use. There are some direct advantages of using templates for the requirements gathering process:

1) *Minimization of errors of omission.* This is somewhat natural to see. Provided the requirements-gathering person is armed with a template that shows all the potential requirements packages that will most likely contain requirements, this template will function practically as a roadmap guiding the gathering process. This template will minimally force the requirements-gathering person to think through the presented packages.

2) *Guidance for the identification of the stakeholder list.* Assume, for instance, that a specific requirements template has a package called regulatory requirements. This is a very strong indicator that the organization quality manager might be a presence in the stakeholder list. This idea can be extended to the corresponding organization functional role for each one of the requirements packages that are included in the organization requirements template.

3) *Reduction of effort.* This is really as simple as it sounds. There might be some boilerplate requirements that need to be present in the vast majority of the

organization technical projects that wouldn't need to be elicited every time a new project is approved.

There is yet another issue that was observed among several organizations: *Lack of Organization-Level Controlling Body.* The PMI advocates that an organization create what is called a PMO (project management office). This office is usually charged with the activities of defining PM processes, best practices, and templates that fit the organization's business. The lack of a centralized body that serves as a gateway of sorts between all projects executed in the past, currently active projects, and project managers prevents the organization from taking advantage of all the benefits of collaboration between past and current projects.

Organizations that are project driven and that don't have this centralized body usually don't offer project managers with the appropriate environment for prevention of the two drivers exposed above: lack of organization process assets and lack of requirements management. Usually what happens in those organizations is that experienced project managers have their own preferred methods, processes, and databases that are used on their own projects, but hardly passed around to other project managers. Less experienced project managers are left to their own devices to survive the cutthroat business of managing complex T&M projects. Ultimately, the overall organization performance on project execution becomes a roller coaster, driven by individual efforts and talents.

Such a management body is not only important for the requirements-gathering activities alone, but also for other areas of project management.

The following diagram summarizes the root-cause analysis performed to identify the most important underlying issues that surface as lack of well-defined requirements.

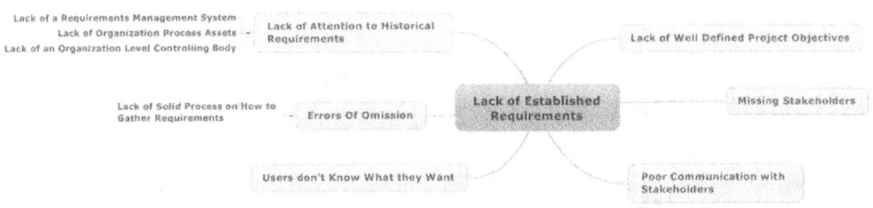

Figure 4. Underlying Issues for Lack of Established Requirements

This chapter provided the root-cause analysis results executed in order to best understand the two main drivers of complex T&M project failure: poor planning and lack of well-defined requirements. The results showed all the main underlying issues that surface as each of the two drivers.

These smaller problems will function as a basis for the remainder of this book. Logic states that since they were identified as the root causes for the two main drivers of project failures, then a framework would present a solution to them. Such a framework is a great candidate to being the solution for the proposed open problem.

The next chapter switches focus to the system integrators, the organizations whose business is to perform T&M systems for clients for profit. They are unquestionably very well equipped to executing T&M systems; however, there are issues clients need to be aware of and make sure are addressed, even when engaging with these organizations.

CHAPTER 2

Outsourcing T&M Projects: Experts Are All We Need, Right?

The short answer to this question is that your organization is certainly much better off by engaging a system integrator to implement the T&M system needed than by attempting to use internal resources. As the old business saying goes, "Focus on the core and outsource the shore." Unless your company has on staff a multitude of engineers who are experienced in putting these systems together as their main job, a system integrator can certainly help in reducing the overall implementation time as well as risk.

These companies' main business is the implementation of T&M systems. They usually have engineers with different backgrounds on staff, who can be pulled together to form a very strong project team. They also should have project managers who are experienced in managing T&M projects and with the pitfalls of integrating these solutions.

However, even though this is usually the case, there are still several T&M projects that end up failing, even when being executed by a competent and reputable system integrator. The challenges in working with such organizations will be the focal point of this chapter, which will present the five most common challenges when engaging an integrator to execute T&M projects, challenges that can actually drive the project to failure. These five challenges will be illustrated with case studies.

Client Trust Initial Requirements Definition to Integrators

As was emphasized previously, the most important driver to failure of T&M systems is lack of well-established requirements. System integrators, as experienced organizations in the business of implementing T&M systems, certainly know that. Integrators' profits stem from their ability to deliver a well-determined scope of work within an original budget estimated and presented to the client in the form of a contract.

It is important to define the term "scope of work." Many people automatically equate scope of work with system requirements, but this is not the case. Scope of work is the actual set of deliverables the integrator will be contractually obligated to provide in exchange for the moneys spelled out in the contract. For example, a given contract might call for two systems to be delivered at the end of the project. Therefore, one of the items that are part of the scope of work for such contract is to deliver these two systems, duplicates of one another. The system requirements document will call out the specifications that both systems will perform. If the contract would be for a single system to be delivered instead, the system requirements document would be the same, however the scope of work, and consequently the price, included in the contract would be very different than the one where two systems are to be delivered.

Another example, a more subtle one, would be training of operators. Training is not something that is captured as part of the system requirements document. However, if called out in the technical proposal, training of operators would be part of the set of deliverables defining the integrator's scope of work for the contract.

Since integrators' profit margins are directly proportional to their ability to complete a contract scope of work within the estimated budget, they are extremely risk averse when creating technical proposals to deliver such systems. This is actually a good thing for the client, since, as seen in a previous chapter, a thorough risk analysis is an extremely important exercise that needs to be executed when implementing T&M systems and one of the first steps toward a successful implementation.

Furthermore, integrators are domain experts in the field of T&M; therefore, it is natural that some clients would feel they should trust the initial requirements definition to them, completely. This, however, is in fact the first pitfall that will be detailed: *Client Trust Initial Requirements Definition to Integrators.*

Let's go through the usual process of engaging an integrator to implement a T&M system and the reason why this can be a potential driver to project failure. A client determines that it needs a T&M system, it shops around, and it invites three integrators to participate in the procurement process.

The integrators, at this point, will start to look for ways to best bind the scope of work, in a manner in which they can actually put a price tag to the effort. This is understandable, as their ability to make a profit is tied to their success in delivering the scope of work for the initial budget captured in the contract.

However, at the same time they are working in defining the requirements for the system that would allow them to define the scope of work and price for the effort, the activity of collecting these requirements is basically a risk for the integrator. Integrators make their money by selling their staff's time. Every hour of staff time they invest in determining requirements for a system they may not be awarded to build is a risk. In our anecdotal example of three integrators participating in the procurement process, all three are investing staff hours to collect the system requirements, but only one will be awarded the contract.

Therefore, they naturally have the focus of making an effort to spend the absolute minimum time they possibly can in order to extract the needed information that would allow them to provide the technical proposal with the scope of work and price. This is not to imply they are cutting corners. Because they are providing a fixed-price proposal to the client, it is also in their best interest to make sure they understand the system requirements enough to not underbid the effort.

However, they will probably focus only on the system features. The system features are what will determine their implementation cost, which will drive the contract price. They will most likely focus on questions such as the following: How many analog inputs will this system need?; Is an analog switch needed? What about a DMM? Does the system need to push data to a database or is a file system enough?

Though these are absolutely questions that need answers as part of the requirements definition process, the requirements elicitation needs to also focus on capturing the business value the system will bring to the organization. Remember the last chapter's discussion about business requirements and business risks as a couple of drivers to project failure? It is almost certain integrators won't drill as deep in the requirements-gathering exercise as to uncover more subtle requirements that are not directly related to their activity of defining the scope of work and price to be included in the technical proposal. This is not because they don't know what they are doing or are cutting corners, but as a basic risk mitigation activity in protecting their business. In the worst case, one can see an integrator running out of business by investing too much staff time in requirements elicitation for T&M systems that end up not being awarded to them.

What this will create is a situation where a T&M system is being defined, a integrator is being engaged to implement it, and the requirements-gathering activity may have missed some key business value requirements or even some subtle technical requirements that would have been uncovered if a more thorough requirements-gathering activity had been executed. This will basically reinforce the *lack of well-established requirements* root cause that was presented in a previous chapter, even though an expert in building T&M system is being engaged to help the client.

A corollary case to what was presented above is the scenario where the client actually pays an integrator to perform the requirements elicitation to kick off the project. This is in fact an excellent practice that should be more often adopted by clients looking to outsource. However, usually, this activity is seen by clients as an expense that is not necessarily adding value to the project. What ends up happening is that a small budget is allocated to this activity; which will force the integrator to prioritize the types of requirements that it will be focused on.

As a rule of thumb, on successful T&M projects, about 25 percent of the project's total hours go toward requirements elicitation and definition activities. If we take an example of a thousand-hour project, which is not that large in today's modern T&M systems, something around 250 hours should be allocated to requirements definition.

In reality, though, I have never seen a client who would be willing to sign up for such an investment up front. Usually, funded requirements-definition projects range between forty and eighty hours in duration.

At this stage, right out of the gate, the system implementation is headed toward a cliff. Usually the outcome for this type of situation is that "new" requirements will be uncovered throughout the project life cycle, as the client is presented with intermediate deliverables. There will be frustrations on both sides, the integrator and the client, as the integrator will see the client as not knowing what she wants, and the client will see the integrator as not being fully committed to the successful implementation of the system.

Every new requirement will have the potential of generating a change order to the contract, which will add cost to the integrator and the client, and the implementation time will suffer. There will be frustration on both sides. The relationship will suffer, and there will be opportunity costs incurred by the client by having the deployment of the system delayed.

One other point to be made is that most of the time, integrators will cover their proposal with what is called technical assumptions. Technical assumptions is a fancy way to say that they either don't know something well enough or didn't have the needed amount of time to fully define something in order to put a price tag on the effort. In other words, the technical assumption defines the assumptions that were used by the integrator in coming up with the scope of work, implementation schedule, and price for the system. The proposals usually have verbiage stating that if one or more technical assumptions are not valid once the contract has been awarded and the implementation is under way, the integrator has the right to modify price, schedule, scope of work, etc.

Obviously, the more gaps there are at the beginning of the project, the more technical assumptions will most likely be invalid. As the client learns more about what is being built by the integrator via intermediate deliverables, these gaps usually start to come to light. Again, in this situation, change orders will be generated to cover such gaps. There is the risk for contractual battles between the client and the integrator in regard to what was and what wasn't included as part of the original proposal. These battles don't add any value to the actual delivered system, most likely add schedule and costs overruns, and, worse, degrade the relationship between client and integrator.

Depending on how the budget for the contract was reserved, in case there isn't enough management reserve put aside to cover the gaps in the technical assumptions, the final system delivered may be close to what was originally proposed due to lack of funding to cover change orders. In this situation, the final system will most likely not capture the maximum business value to the client and the integrator may end up with a money-losing project, a lose-lose relationship.

Figure 5 summarizes the issues that are the by-product of this first challenge.

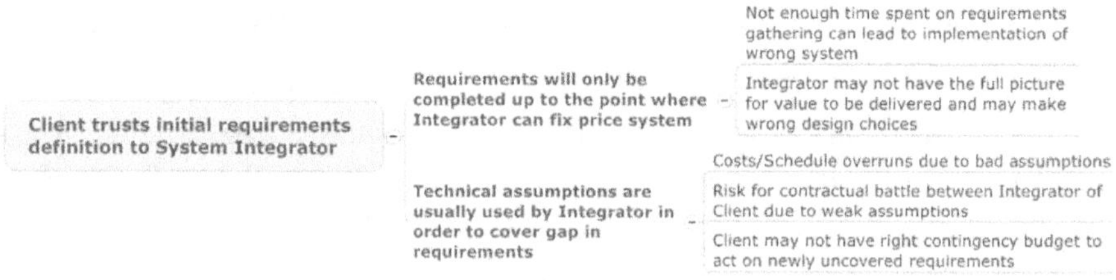

Figure 5. Issues with Client Trusting Initial Requirements Gathering to Integrator

First Challenge: Case Study

A client hired an integrator to build a full T&M system as well as several interface test adapters (ITAs) to test functional boards that when assembled would make up the client's new product being launched to the market. Each ITA would be used to interface one of the DUTs (devices under test) back to the main T&M system.

Integrator was part of a competitive procurement process with a few other integrators. There was no funding from the client for an initial requirements-gathering activity, and the client trusted the integrators to perform their own requirements gathering as part of their proposal submittal process.

Client provided DUTs schematics and an initial set of test specifications for each one of the DUTs. The information set was considered to be more comprehensive than clients usually provide integrators at early stages of T&M projects, which boosted the confidence that a high-level requirements phase could be executed in order for a fixed-price proposal to be generated.

And so it was, and the integrator was awarded the contract. As the project was under way, a new piece of information came to light. The client was counting on having the system deployed in a foreign country by a certain date not too distant in the future. This was extremely important to the client because it would reduce the manufacturing costs of the product by a significant factor. There were different departments that were counting on this to happen by the target date. In summary, the business value for this T&M system was much, much greater than originally known. This put an extra pressure on the integrator, as now the schedule became extremely firm, and no margin for error was allowed.

Once acceptance testing started at the integrator's shop floor, several of the DUTs were not passing the tests. As it turned out, there were couplings between the T&M system components and the DUTs that were causing the tests to fail. After further in-depth analysis, it was learned that the DUTs were much more difficult to test than the high-level requirements gathering executed by the integrator had suggested.

The tester needed to be redesigned to account for this extra complication layer in the DUTs. Cost and schedule overruns resulted, which pushed the final deployment date past the important target the client was counting on.

Client Does Back of Napkin Requirements Definition

Even though this may sound like a step up from having the client fully trusting the integrator to perform requirements elicitation, it can actually be as dangerous as the former scenario.

This scenario is when the client does have internal resources that can be devoted to an initial requirements-gathering exercise. Though this may sound appealing as there is nobody better to understand the pain points and the areas of value that need to be captured by the T&M systems as the client herself, there are some underlying challenges that need to be addressed.

The first one is the fact that the client may not be as well versed in performing a thorough requirements analysis for T&M systems as an expert from the field would be. Refer back to the previous chapter where it was presented that whoever is gathering requirements needs to make sure to pay equal attention to both the technical and business requirements.

This drives errors of omission, as described in the previous challenge. Errors of omission will in turn drive change orders by the integrator, which will drive schedule and cost overruns, much like in the case of the previous challenge described in this chapter.

Also, it was mentioned in the previous chapter that it is usually extremely difficult to find resources that can execute a well-balanced business and technical requirements-gathering activity. It is important to recognize potential gaps in the organization's resource pool to make sure the company is not falling into the traps exposed in the previous chapter around the challenges in gathering well-formed requirements.

One other fact that is usually present is that internal resources rarely can dedicate themselves full time to the requirements-definition exercise. They usually have their day-to-day job they still need to perform, and are asked to execute this other activity on top of that job. As it was presented, planning and preparation for the T&M project is the most important driver for project success. Shouldn't such important activity deserve a full-time resource(s) working on it?

Another fact that is fairly common is for internal resources not to have in-depth knowledge of the type of information an integrator usually needs from clients in order to keep headed in the right direction. Unless the internal resource has worked for a system integrator in the past or has lots of experience engaging integrators as part of her job, the integrator will most likely have to direct this internal resource on the type of information that it needs in order to successfully quote and execute the project.

This knowledge on how the integration business works is extremely important when performing the initial risk analysis for the T&M project. If an internal resource doesn't necessarily know what drives risk for integrators, it may not properly prioritize requirements that could have been dropped to reduce the overall proposal price with minimum impact to the final solution.

Sometimes, the execution of prototyping activities to further understand risk areas is another extremely valuable activity. The understanding on how risks affect not only the T&M project as a whole, but also the integrator implementation, is fundamental for the definition as to whether or not prototyping is needed and to define exactly what is to be prototyped and how.

One may argue that the integrator can potentially suggest this type of trade-off and prototyping activities, but it is usually difficult for that to happen unless the integrator is very familiar with the project and client and has spent a considerable amount of time understanding the system to be built, how it will affect the client's business, pain points to the client, etc. As it was seen at the first challenge described in this chapter, integrators don't usually have that time available.

When this is the case, the relationship falls back into the previous challenge described, as the integrator's focus, as it was seen, is in making sure it is delivering the contracted scope of work, and not necessarily aligning the T&M implementation to maximize its business value to the client. The integrator will most likely use several technical assumptions to cover the gray areas of knowledge in the requirements set and will probably provide a much higher proposed price for the system to cover the risks.

In this situation, one suggestion would be for the client to consider having an expert who can direct the internal available resources in the right direction and make sure the facts above don't push the planning and preparation toward failure. This expert would, ideally, have a thorough understanding of the system integration business, have deep technical skills in the technology available to implement the technical solution, and, ultimately, have business savvy to understand the business value the system will bring to the client.

Figure 6 summarizes this challenge and its consequences.

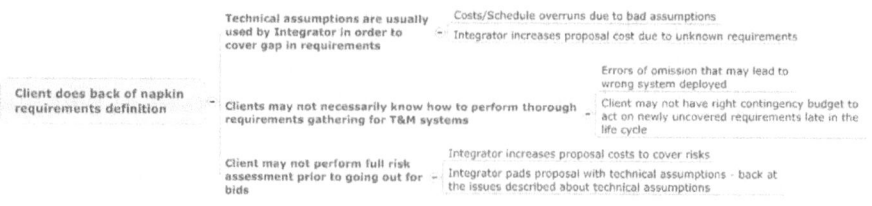

Figure 6. Issues with Client Doing
Back of Napkin Requirements Gathering

Second Challenge: Case Study

Client hired integrator to create a T&M system that would test the new generation product to be launched a few months later.

The DUT required a custom electronics board to be designed and implemented in order to properly interface the device under test with the T&M system instrumentation.

Client didn't fully understand risks on T&M systems and decided to reduce the scope of work by integrator to just the instrumentation and the software, in order to reduce overall contractual costs. The client was going to take on the design and implementation of the custom electronics board by using an internal resource.

Integrator included in the contract that its scope of work would be verified and accepted by the execution of an acceptance test procedure that would not include the client's designed interface board. Client didn't fully understand this contractual verbiage.

Integrator completed its scope of work and required contract closing as the contract was fulfilled. Client withheld contract closing until custom interface board was completed. Incidentally, the integration of the board didn't work as planned by client as integrator was given bad requirements for the acquisition software.

A contractual battle started as the client was refusing to close out the contract because the overall system wasn't working. Integrator was right in requesting contract closing as it has fulfilled its obligations. However, the fact of the matter is that the T&M system implementation got delayed, which pushed the client's NPI by a few months.

Over the Fence Mentality

This situation happens when the client believes that since an expert company is being hired to execute the T&M system, the system integrator will take care of everything with no or minimum interaction with client internal resources until it is time to deploy the system.

In this described scenario, not only the requirements gathering is given to the integrator, but there is usually minimum interaction throughout the project life cycle. It is important to make the distinction here that when it is said that there is minimum interaction with the client throughout the project life cycle, the intention is to say that there is minimum technical and business exchange between the client and the integrator. Even though the integrator's project management process may call out milestone meetings along the course of the project, in this scenario, the client either doesn't have technical resources to make sure the proposed solution by integrator is in line with what is needed, or it doesn't have available resources that can devote the time to make sure the client's business value is being implemented.

In this situation, the client usually flies through the milestone checkpoints and raises a flag much later in the project life cycle, usually at deployment time, when the delivered features "are not what they thought they would be."

Another issue that is usually noticed in this scenario happens around the information flow between client and integrator. Unless there is a main point of contact on the client's side who understands what type of information needs to be passed on to the integrator and the potential impact of not passing on that information will create, usually the communication between client and integrator suffers. Once the communication suffers, the project is headed to failure.

As a trivial example to the paragraph above, consider a scenario where the integrator is working on developing a T&M system to test a device that is under development. Assume now the common scenario where the DUT development team updates the DUT firmware. This obviously can cause an impact on the development of the T&M system depending on its design and how it is communicating with the DUT. If the point of contact on the client's side doesn't understand that this needs to be communicated to the integrator, the project team will continue development of the T&M system as if nothing has happened to the DUT. This can lead to an unpleasant surprise at verification time when the integrator can spin its wheels trying to debug the system, without knowing the DUT is actually different than the one the system was designed to interface with. This will lead to cost and schedule overruns.

This was mentioned earlier, but it is worth reinforcing. It is in the client's best interest to make sure the final T&M system delivered by the integrator is aligned to maximize its business value to the client. Good integrators will make sure the contract is fulfilled and the T&M system that was originally contracted is delivered. However, as mentioned in the previous chapter, it is sometimes very difficult to predict all features and interfaces a T&M system needs to implement at very early stages of its project life cycle. As the system design and implementation is under way, and parameters from other areas that are related to how the system will be utilized by the organization become clearer, sometimes there is a need for tweaks and small redirections. The integrator who is left to her own devices won't have the needed visibility to make those adjustments, so an opportunity to capture value may be missed.

Figure 7 summarizes the consequences of this scenario.

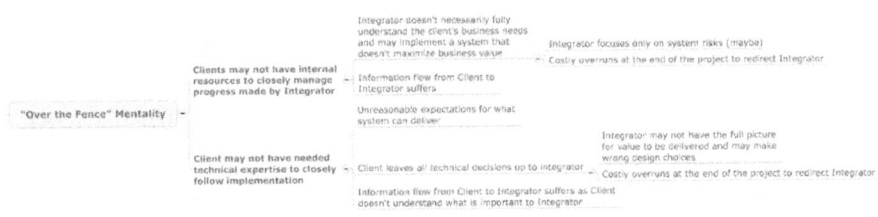

Figure 7. Issues with Over the Fence Mentality

Third Challenge: Case Study

Client hires integrator to implement a complete data acquisition system on a state-of-the-art building. All clients' resources involved with integrator had background in construction and virtually no knowledge in T&M.

Client, in several instances, failed to communicate key information very important to the integrator's design. Also, it became very difficult for integrator to communicate the design intention and even how the system was to be utilized until the actual system was built.

At deployment of some components, client noticed very important gaps in what the expectation for the full system feature set was versus what was being implemented by the system.

A full module was missing. The completion schedule was approaching fast. The integrator had to scramble to provide a quotation for the missing features in the form of a change order. By that time, the management reserves for the whole project were running very low and there was virtually no budget left for the implementation of the new module. Very long negotiations took place, which added no value to the actual project and pushed the final deployment schedule several months out.

T&M System Implemented in Parallel to DUT Design

One other scenario that is fairly common in the T&M industry is when, in support of a client's NPI, a T&M system needs to be designed and built prior to the actual device under test being designed and built.

One may question the feasibility of the approach; however, the fact of the matter is that the vast majority of the NPI schedules always assume the test system will be ready to be used as soon as the product development is complete. This sounds like an impossible task and a recipe for disaster as it violates the root cause for T&M project failure: lack of well-established requirements.

In fact, it can be an impossible task and a recipe for disaster, depending on how the project gets executed. As it will be seen in a later chapter presenting the TMPM framework, there is a solution for this problem. However, this is usually not how the situation is tackled.

Historically, integrators have been applying a typical waterfall approach to managing T&M projects. As seen in the introduction, the concept of waterfall development lies in the idea that the overall project can be broken down into different phases and that work needs to be brought to completion in one phase in order for work on the next phase to be initiated. For instance, all requirements need to be fully defined before system design can start.

Waterfall development thus emphasizes more time spent up front in the project life cycle, highlighting the importance of requirements and design documentation. It is easy to see why integrators favor this methodology. Since they are on the hook to provide clients with fixed-price proposals for T&M systems, the only way for them to have a level of confidence that the risks are somewhat known and under control and that they have a shot at being profitable is to make sure the requirements are known to a somewhat high level of certainty.

One can see the conundrum of this scenario. How can the T&M system requirements be defined and well known if even the DUT requirements themselves are still in a high state of flux?

What ends up happening is that the integrator will follow its usual waterfall model by using the information that is available about the DUT at the time of proposal generation. It will use technical assumptions to make sure it is contractually covered in the event of changes to the DUT. The proposal budget is taken to whoever is the economical buyer on the client's side, a T&M budget is set aside, and the contract is put in place.

Well, changes to the DUT not only are expected, but are certain to occur as the product development process is still under way. A project being executed under this scenario needs to be managed extremely well, not only on the integrator side, but also on the client side. However, one first issue that will most likely happen is the fact that the original budget that was put in place for the T&M system will most likely not be accurate due to the fluidity of development. Depending on the depth of the changes in the DUT, the T&M system may need to change significantly to match the new revision of the DUT. This always carries cost and schedule impacts.

Using the last sentence as a hook to the next argument, the original implementation schedule that was included in the proposal will most likely not hold, for the very reasons that were presented above. Now the expectations need to be realigned, not only with the client's main point of contact who is managing the integrator, but throughout several other departments in the client organization that were assuming the T&M system original schedule to be gospel. This communication needs to be very tactfully managed by both the integrator and the client's main point of contact who is managing the integrator.

One other fact that is very often overlooked by both client and integrator is that usually there is a lot of value in having the T&M system partially delivered to the client and in a configuration where the product development team can use it to execute characterization exercises. The T&M system has instrumentation that usually can used by the client R&D department to collect data that will help in making design decisions about the DUT.

If a pure waterfall development methodology is used, the integrator will most likely object to having these intermediate deliverables to the client, as they can be very disturbing to the actual flow of development of the T&M system. They most likely will add to the integrator's internal costs and will create an impact on the overall deployment schedule.

These issues are very difficult to address by the existing frameworks and the typical way integrators execute T&M systems. Later chapters show how the TMPM framework proposes to address them.

Figure 8 summarizes the consequences of this challenge.

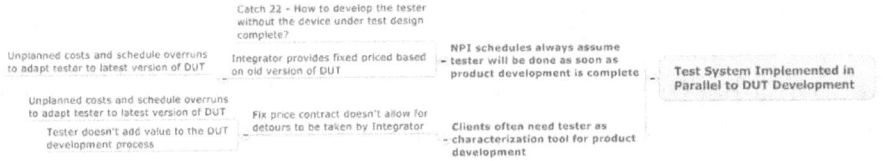

Figure 8. Issues with Test Being Developed in
Parallel to DUT Development

Fourth Challenge: Case Study

Client hires integrator to develop a T&M system and ITAs for several functional boards that together form a new product to be launched to the market a few months later.

All of the functional boards are on very early stages of development. Client's NPI schedule assumes the ITAs will be available as soon as each of the functional boards' development is complete.

Client shared NPI schedule for the multiple releases of the many functional boards and the overall NPI schedule milestones. Integrator, in predicting the scenario that was described above, included provisions in the proposal to execute multiple revisions on the ITAs, as new revisions of functional boards became available.

Client ended up deciding not to accept integrator's proposal for multiple revisions to reduce T&M costs, as it didn't have an internal T&M champion who understood the consequences of what client was trying to accomplish.

No staged delivery was implemented, and therefore the client's R&D department didn't have available a tool with which to execute characterization exercises. R&D engineers were forced to create their own bench test setups, which ended up sinking precious development time they could have been applying to the DUT itself in support of the NPI.

Furthermore, integrator had to wait several weeks for the official go-ahead by the client, signaling that the first functional boards had been released. These several weeks could have been used to develop features of the ITA that would be common among the multiple revisions of the DUT and shorten the overall T&M implementation schedule.

T&M Budget Defined Before Feature Set Is Complete

This scenario is somewhat similar to the one presented above where the DUT design is done in parallel to the T&M implementation. However, it was included here in order to highlight the fact that this situation where the client needs to define the budget for the T&M system prior to its functionality set to be fully defined is not a situation that is exclusive to a client's NPI processes.

Several clients define the business need for a T&M system and need to come up with some dollar figure in order to justify that capital expense. Usually, a procurement process is started and integrators invited to provide technical proposals for implementation. Very often, at that early stage, the full feature set for what the T&M system will implement is not necessarily flushed out. It is not uncommon for clients to request an ROM (rough order of magnitude) estimate in order for capital budget to be secured.

The integrators will work with the information that is available at the time and come up with a fixed-price proposal or an ROM that covers the known features as well as include technical assumptions to cover risks and unknown features.

The best technical proposal/ROM approach is selected. At that point, the capital budget is reserved and maybe the project is started immediately by awarding an integrator the contract, or it starts at a later date. Regardless of when the project starts, the fact of the matter is that now there is a fixed budget set to implement a system where its feature set is not fully determined.

New requirements on these situations will always come to light once the project is under way. As seen previously, the integrator will try, to the best of its ability, to stick to whatever scope of work was attached to that project budget. There will be an inherent disinterest from the integrator's point of view to act and implement new requirements. Or its change orders will be put in place as a mechanism to make sure the new scope is properly funded. However, as it was seen, there usually isn't budget available in addition to the capital budget that was reserved for the project when the feature set was still undefined.

This causes all sorts of problems and one of the three following consequences: integrator will offer to drop existing scope and use corresponding funding to the implementation of the new requirements; new requirements will not be implemented; or there will be a contractual battle between client and integrator in case the proposal verbiage is somewhat ambiguous in relation to requirements and scope of work.

Regardless of which of the three actually happen, there will be dissatisfaction and non-value-added activities happening between the two organizations that will most likely cause cost and/or schedule overruns. And it is almost guaranteed that the implemented system will fail to maximize its business value to the client.

As it will be seen in later chapters, the TMPM framework has a proposed solution for this very common scenario in the T&M industry.

Figure 9 shows the consequences of this challenge.

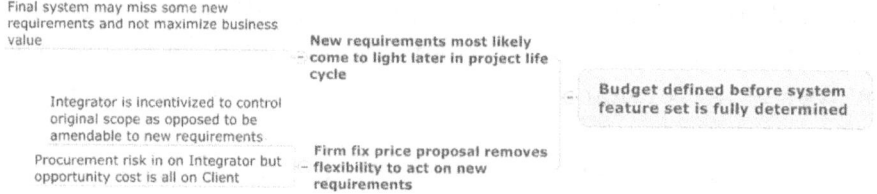

Figure 9. Issues with Budget Defined Before Feature Set Complete

Fifth Challenge: Case Study

Client hired integrator to execute a high-speed data acquisition system that would require very advanced custom electronics to interface the device under test with the system instrumentation.

Client executed a very thorough initial requirements elicitation phase, prior to engaging integrators for proposal submittal. As it turns out, client's initial requirements had instrumentation specifications that couldn't be achieved with existing off-the-shelf instruments.

Integrator proposed an initial discovery phase where the high-level design would be put together and a theoretical set of specifications derived so a gap analysis between the original specification from the client and the one that could be achieved with off-the-shelf instrumentation could be determined.

Next, once the theoretical gap was determined, integrator and client agreed that integrator was to execute a small project composed of a proof of concept. The goal for this exercise was to determine the practical specification gap between the proposed solution and the initial specification set. At that point, the client would make a decision as to whether to continue down the path of full implementation for the system or to cancel the project.

As it turned out, this project, though extremely challenging from the technical standpoint, ended up as a low-risk one for both client and integrator.

Can all the issues that were presented in this chapter be pinned as the integrator's fault? Hardly. Are they the client's fault then? Also not true. As it was mentioned in the introduction of this book, today's T&M systems have became much, much more complex than they were several years ago. The pace of technology advancements pushed end products to become true engineering marvels. As such, the corresponding T&M systems to support such products are also now exponentially more difficult to implement than the ones supporting much simpler devices under test.

Furthermore, technology advancements also reached data-acquisition hardware devices that are now available as off-the-shelf components. What can be done now with off-the-shelf components not so long ago required that a product development project for a custom electronic device be created. The breadth of tools available to engineers and scientists today allow them to unleash their creativity in solving the most difficult problems. As a direct consequence, the requirements for T&M systems of today have transformed such projects into very complex projects. As it was seen in previous chapters, complex projects need extra levels of planning and requirements-elicitation processes in order to be successful.

The conventional model that used to work, where a client would engage an integrator as the domain expert in the T&M field and simply hand the project off as a guarantee of success, is no longer valid. The path to success when engaging an integrator to execute a T&M system these days requires that the client understand that a true partnership needs to be formed with the integrator.

It is in the client's best interest that the integrator has an uneventful project implementation, as that is the guarantee of an on-time high-quality system being delivered. For that to be the case, the client needs to better prepare prior to engaging an integrator to execute the T&M system as well as to support and monitor the integrator throughout the entire project life cycle to make sure the integrator is marching in the right direction.

Conversely, the integrator needs to make sure it fully understands and is aligned with the T&M system business value provided to the client. It is in the integrator's best interest to raise a flag when/if it recognizes that the current contract as stated would prevent the organization from providing the most value to the client. As in the case of defects that cost much less to fix when uncovered early in the project life cycle, it is much easier to deal with these types of contractual issues when the project is not going in a critical phase and everybody is scrambling to deliver it.

Usually, at early stages of the project life cycle, the integrator has much more flexibility in moving funds around and reprioritizing tasks, as those funds haven't been burned as part of the project execution. One universal truth that needs to be understood by both client and integrator is that a risk won't just go away if it is ignored. Making sure risks are always being captured doesn't mean the people involved in the project are being negative or pessimistic about its outcome. On the contrary, the project team—and when I say project team here I mean all project stakeholders from both integrator and client—that is constantly bringing up risks to be captured as part of the risk registry is the one that usually sees the best overall project outcome for both integrator and client.

As it was seen in the first two case studies presented in this chapter, the client would have been much better off by executing a thorough requirements elicitation exercise prior to engaging the integrators in the procurement process. If the client didn't have internal resources available to perform such activity, it would have been much better to have engaged an external consultant experienced in the T&M industry and in how system integrators operate, in order to make sure things were much more advanced and would raise the odds of success for the integrator.

For the third use case, higher odds for project success would have be obtained if the client had, either as part of its internal staff or as an external consultant, an expert in T&M to function as the link between the integrator and the client's construction stakeholders. This person would make sure the information needed by the integrator would properly flow from the client, align the technical implementation by the integrator with the high-level business requirements, and make sure all expected features by the client's stakeholders were being included within the scope of work to be delivered by the integrator.

As was presented by the third challenge earlier in this chapter, even if the integrator has a very experienced project team and project manager in charge of the T&M project, the over the fence mentality always ends up leaving gaps that usually only come to light at late stages in the project life cycle.

The fourth and fifth case studies have similarities in the fact that an integrator needed to be engaged by the client prior to the T&M feature set to be fully determined. This is a fairly common scenario in the industry.

They were presented to illustrate the fact that the conventional model integrators and clients engaged are not conducive to a successful project under this condition. The fifth case study presented an engagement model that reduces the overall risk to both integrator and client, as well as maximizes the business value for the final delivered system.

Opportunity costs for a delayed product launch are the highest cost an organization will bear with today's speed-to-market requirements, which can actually knock companies out of business. It is important to keep in mind that T&M systems will always be a key component for any NPI effort. A new product can't be launched to market without being thoroughly tested. As such, the same amount of planning and preparation that is devoted to the activity of designing and implementing the final product that will be launched needs to be applied to the T&M system that will support the product launch.

Any perceived costs in proper planning and preparation for the T&M effort is money very well invested in the success of the NPI. As it was seen above, money spent up front is not necessarily an added cost to the overall T&M activity, but a redistribution of that cost and potentially a cost reducer. It is a known fact that defects found at earlier stages of a project life cycle are much cheaper to address than the same defects uncovered at later stages. As such, it is easier to understand that every dollar invested on planning and preparation for the T&M project activity, even when engaging an integrator to implement the system, will save the client many dollars later in both saved change orders during project execution, but chiefly, in saved opportunity costs by speeding up the NPI.

Solving the issues presented in this chapter brings several benefits to both integrator and client: overall project risk reduction to both integrator and client, potential reduction of the price of the proposed solutions by integrators, reduction of errors of omission, and lowering of odds for cost and schedule overruns.

All of these features bring to the client:
- Reduced cost of ownership for the T&M system
- Reduced opportunity cost due to delayed NPIs
- Allows client to use T&M system in support of product development activities
- Allows integrator to more efficiently become an extension of the client's internal staff
- Maximizes business value per cost

From the integrator's perspective:
- Reduces the overall risk on fixed-price projects
- Drives more predictable gross margins for the business
- Provides opportunities for cost reductions due to lower risky contracts, which can make the integrator more competitive
- Increases flexibility to make client successful without corrosion of project gross margins
- Provides overall higher value to clients, which will drive more repetitive business

The next natural step is to analyze the existing frameworks utilized for management of T&M projects. The following chapter will present the areas of PMI's project management framework and INCOSE's systems engineering framework relevant to the issues identified for T&M project failure. The chapter after next will then expand this analysis to cross-reference those frameworks with the challenges in working with an integrator, as listed in this chapter.

CHAPTER 3:
Existing Project Execution Frameworks

Now that the main drivers of T&M project failure have been properly broken down into their corresponding root-cause items and a thorough analysis was performed in regard to utilization of system integration companies in the implementation of such systems, the next natural step is to perform an analysis of the existing frameworks that can be applied to the execution of these projects.

The objective of this analysis is therefore to understand if there are any gaps in the existing frameworks that prevent clients and system integrators from addressing the identified root causes in the execution of their T&M projects. The rationale is straightforward: if there are no gaps in the existing frameworks then the issue might be enforcing the proper project execution under the umbrella of those frameworks. Should that be the case, the whole solution might boil down to a matter of training in some cases, and of better monitoring and controlling the execution activities.

This could suggest that a well-designed PMO (project management office) could potentially do the trick. In summary, a PMO is an organization body that executes one or more of the following functions:

1) Responsibility for the creation and implementation of the organization's project management process
2) Responsibility to regulate the activities of project managers for compliance with the organization's project management process
3) Make sure the selected projects provide good alignment with the organization's strategy and is capturing the most business value

4) Functions as part of the change control board in order to analyze and provide approval or rejection for requested project changes along the course of the project life cycle
5) Responsibility for the training and professional development of the organization's project managers
6) Some PMOs are also responsible for the execution of the project management activities in the low level

As mentioned previously, should the existing frameworks be free of any gaps in relation to the identified root causes for the two main drivers of project failure, then our attention should be focused on making sure a PMO is in place and on how it is implemented in organizations. At that point, it would make sense to search for root causes in the implementation of the PMOs that lead to the root causes of technical project failure.

Conversely, if the existing frameworks do show gaps in relation to the identified root causes for the two main drivers of project failure, then the potential solution to the problem will come via bridging those gaps into a framework that addresses the identified issues that drive project failure.

This chapter will perform this analysis on two of the main frameworks that are widely utilized in the industry for the management of T&M projects: the Project Management Institute's PMBOK® and INCOSE's systems engineering frameworks.

These underlying issues will serve as the basis for the analysis of the existing frameworks. Only the topics relevant to these issues pertaining to each of the two frameworks will be presented, and the strengths and weaknesses analysis on each of those topics will follow.

The PMI PMBOK® Framework

The first framework that will be analyzed is what we will be calling here the PMBOK® framework. This framework is the result of years of work by the project management community, organized and led by the Project Management Institute. The summary of the knowledge was concentrated in what is known as the PMBOK®, or Project Management Body of Knowledge.

The current version of the PMBOK® provides best practices around project management in the format of forty-two processes and their corresponding inputs and outputs. These processes were organized in what PMI calls process groups. The entire body of knowledge was grouped into five process groups: Initiating, Planning, Executing, Monitoring, and Controlling and Closing.

Figure 10 shows the five process groups as well as their interaction with each other. The two main project failure drivers and their root-cause items are part of both the initiating and planning process groups.

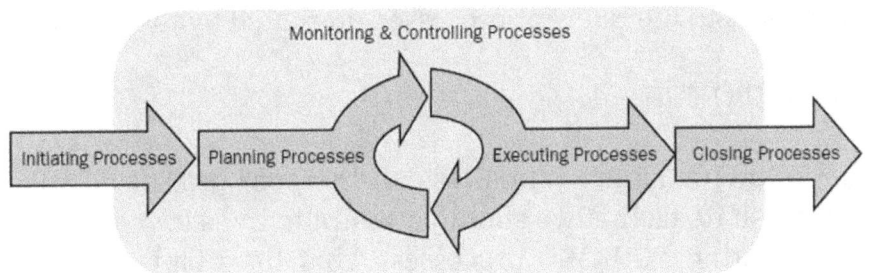

Figure 10. PMBOK® Framework

The entire PMBOK® thus is a collection of processes that can be understood as the activities that need to be executed by a project manager in order to successfully manage a project. The framework presents all of these processes and suggests that the project manager make sure that only the needed processes for a given project be selected in order to avoid wasted effort.

As mentioned previously, one main characteristic of this framework is that it is designed to be applicable in any type of project, not necessarily T&M projects. The PMBOK® defines a project as a "temporary endeavor undertaken to create an unique product, service, or result." Therefore, any endeavor that meets this definition, in theory, can be managed by the application of either a subset or all forty-two processes presented by the PMBOK®.

Based on this initial high-level discussion, one can see that it should be reasonably straightforward to map a single or multiple PMBOK® process(es) to each one of the identified root causes presented in a previous chapter of this book. As mentioned previously, should the PMBOK® provide at a minimum a single process that addresses each one of the root causes, then the PM framework is a good candidate to become the framework of choice for the management and execution of T&M projects.

The following subsections will thus explore each one of the main drivers for T&M project failure and their corresponding root causes to the relevant processes presented by the PMBOK®.

Poor Project Planning

The following table summarizes all the poor project planning root-cause items that we should attempt to trace to one or multiple existing PMBOK® processes. This list displays all the items captured by the diagram presented early in this chapter.

ROOT-CAUSE RESULT FOR POOR PROJECT PLANNING
1 Lack of Established Project Objectives
1.1 Poor Communication with Stakeholders
1.2 Missing Stakeholders

1.3 Lack of High-Level Scope of Work
1.4 Lack of Acceptance Criteria
2 Poor Risk Identification
2.1 Unbalanced Business and Technical Risk Identification
2.2 Not All Appropriate Stakeholders Involved in Risk Identification
2.3 Project Manager Tries to Do It by Herself
2.4 Risk Register Is Not Made a Living Document
2.5 No Formal Risk Identification Process Done as Part of the Precontract Phase by Service Organizations
2.6 Lack of Understanding of the Importance of Risk Identification
3 Lack of Validated Assumptions
4 Lack of Stakeholder Buy-In to the Plan

Table 1. Identified Root-Cause Items for Poor Project Planning

As stated previously, the PM framework is divided into five process groups. Incidentally, all the identified root causes for project failure, repeated for convenience in the table above, should be addressed by either initiating or planning process groups.

The following sections present a summary of those process groups and how they would potentially address the poor planning underlying issues. It is important to keep in mind that the text being presented under the following sections assumes that the conventional PMBOK® framework, with a single project manager, is being utilized.

Initiating Process Group

The very first output the PMBOK® identifies as a result of the Initiating process group is a document called the project charter. The project charter is a very important document that officially authorizes the start of the project. It assigns a project manager to the corresponding project and gives this project manager the needed authority to start the activities to manage the project.

In summary, the project charter presents the business case that justifies the creation of the project and its value to the organization as well as the problem the project must solve or, in other words, the high-level scope of work or the business value the organization is attempting to generate.

The project charter, as defined by the PMBOK®, is one of the outputs of the Initiating process group, the others being the Stakeholder Registry and the Stakeholder Management Strategy. What this means is that the processes that are part of the Initiating process group will be applied in conjunction by the project manager to produce three output artifacts that will serve as inputs to the Planning process group.

One of the biggest mistakes made by T&M project managers is to overlook the importance of the project charter. T&M project managers often believe this nontechnical document is nothing but a non-value-added formality; however, this document actually can and should function as a checklist to the project manager in order for her to make sure very important information that needs to be collected prior to the project being authorized is properly compiled.

The PMBOK® defines a process called Develop Project Charter that calls out the need for the project charter to be created. There isn't much more information provided by the PMBOK® other than the inputs that should be provided to this process in order for the project charter to be created. As defined, these inputs are Project Statement of Work, Business Case, Contract, Enterprise Environmental Factors, and Organization Process Assets.

What is called Enterprise Environmental Factors by the PMBOK® is defined as "both internal and external environmental factors that surround or influence a project's success." This includes the following items, to cite a few:

- Organizational culture and structure
- Organizational human resources
- Project management information system in place for the organization
- Factors external to the organization scope, such as marketplace conditions, government and industry standards, etc.

What this means in practical terms is that the project manager should take into consideration factors that not necessarily belong to the project per se, but are present in the surrounding environment and will add to the overall project landscape.

The Organization Process Assets can be summarized as anything that already exists as part of the organization that can be leveraged in the management of the current project, such as project management process, documentation templates, project life cycle, etc.

The Develop Project Charter, as defined by the PMBOK®, is summarized in figure 11.

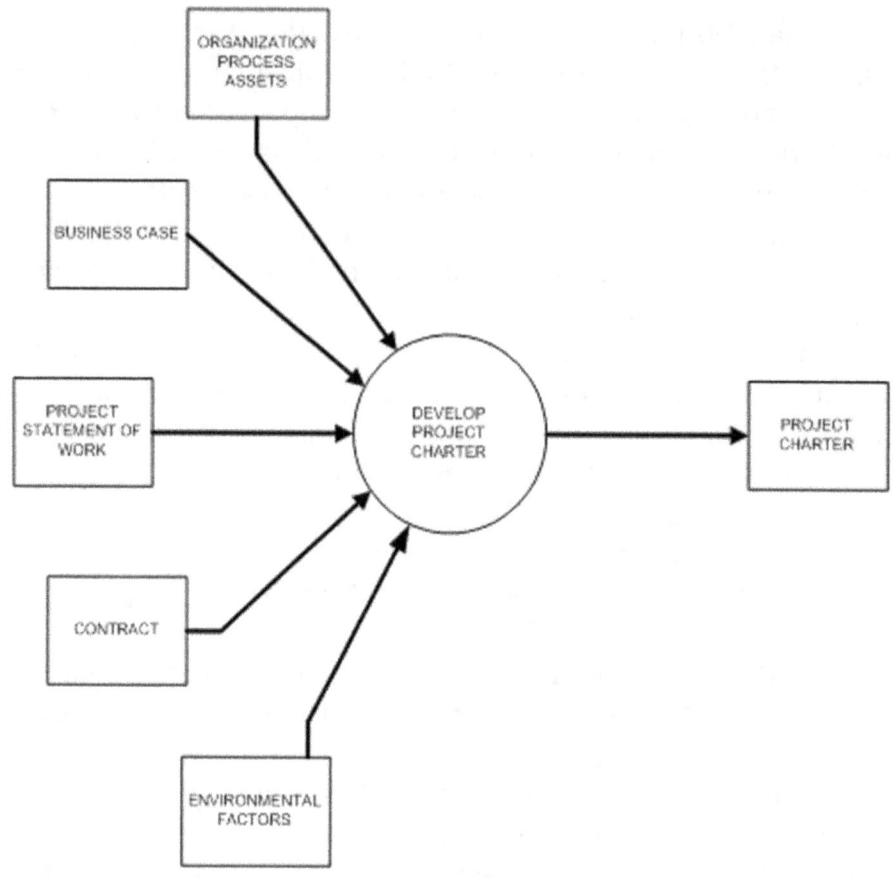

Figure 11. Develop Project Charter Process

Now, let's take a step back and look for potential intersections between the Develop Project Charter PMBOK® process and our root-cause analysis for poor project planning.

Referring back to the identified underlying issues for poor project planning, one can identify two of those issues as being potentially related to the Develop Project Charter process: Lack of High-Level Scope of Work and Lack of Validated Assumptions. As one can notice based on figure 11, the project statement of work is one of the inputs to the Develop

Project Charter process. A good statement of work should actually include validated assumptions as well as a high-level scope of work. With that said, as we have made the case in a previous chapter to show how these two artifacts are drivers of poor project planning, what this indicates is that there is a fundamental problem rooted even before the project charter is created.

This indicates that the project charter might be created under the old saying "garbage in, garbage out"; regardless how much due diligence is invested in the creation of the project charter, this artifact might be missing some fundamental information that didn't get captured by one of its inputs, the project statement of work.

Even though this can't be considered as being a gap in the PM framework per se, this will set the project up for failure as all the planning activity will be based off incomplete or wrong information. Since the PMBOK® doesn't currently have an answer for this issue, we will identify it as a gap in the framework that needs addressing, for the purpose of our analysis.

Still going through the Initiating process group, the other two outputs of it, as previously mentioned, are the Stakeholder Register and the Stakeholder Management Strategy. The Initiating process group has another process that is applicable to generating these two outputs, namely Identify Stakeholders.

As defined by the PMBOK®, the Identify Stakeholders is the process of identifying all people or organizations affected by the project and documenting relevant information regarding their interests, involvement, and impact on project success.

The Identify Stakeholders, as defined by the PMBOK®, is summarized in figure 12.

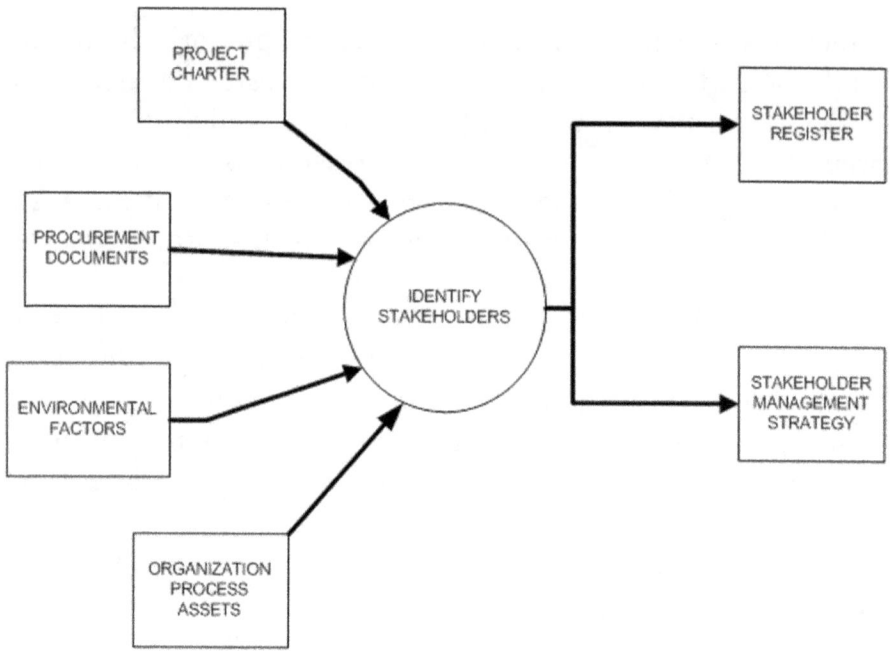

Figure 12. Identify Stakeholder Process

Though this methodology is great in order for the project manager to capture and compile basic information about the stakeholders, the PMBOK® only provides high-level information on the need for this process, as well as its inputs and outputs. There is no detailed methodology and/or best practices showing how a project manager can successfully execute this activity. The project management literature provides some recommended tools and techniques to be applied to this activity; however, there is a gap in tools and techniques that are geared toward T&M projects.

The main issue here is error of omission on the Stakeholder Register. Sometimes, especially in the event of T&M projects, it is not as straightforward to determine what stakeholders should be included as part of the Stakeholder Register. There might be some hidden dependencies between the multiple disciplines that are involved in the technical aspects of the project as well as between those disciplines and the business aspects that don't become clear by a simple exercise of listing potential stakeholders. As explored in previous chapters, failure in identifying a stakeholder is disastrous on execution of T&M projects; therefore, the project manager must pay attention to this activity as needed.

Now, let's analyze the other output of this process: the Stakeholder Management Strategy. As defined by the PMBOK®, it defines an approach to increase the support and minimize negative impacts of stakeholders throughout the entire project life cycle. It includes elements such as:

- Key stakeholders who can significantly affect the project
- Level of participation in the project desired for each identified stakeholder
- Stakeholder groups and their management (as groups)

Usually, the most common way to present the Stakeholder Management Strategy is via an analysis matrix. An example of a stakeholders analysis matrix is seen in table 2.

Stakeholder Name	Interest in the Project	Impact	Strategies for Gaining Support

Table 2. Stakeholders Analysis Matrix

Though the Stakeholder Management Strategy is a great tool to help the project manager in planning communications and well as managing stakeholder expectations, it doesn't really solve the issue of multiple backgrounds between stakeholders, as mentioned in previous chapters. This will therefore be seen as a framework gap.

One last comment about this process: it is important to note that procurement documents are important inputs to the Identify Stakeholders process. Also, it is natural to understand that the make-or-buy decision that will potentially generate the procurements documents will only be performed much later in the project life cycle. Specifically for T&M projects, this may include the engagement of a system integrator. As seen in a previous chapter, one of the presented challenges in engaging an integrator was the scenario when the budget needs to be determined prior to the system feature set being determined.

This process assumes procurement documents will be in place, which adds fuel to the fire of engaging an integrator too early in the overall project life cycle.

This indicates that the Identify Stakeholders process should be a living process, at a minimum through the end of the project planning. The end of the project planning will determine the creation of the project plan, as will be seen in later sections. The project plan should contain as a component the procurements management plan, which indicates the make-or-buy decision that needs to be executed prior to the creation of the plan.

It is good practice for the project manager to take one more pass at the Identify Stakeholders process at the end of the project planning activities, so that any potential new stakeholder that wasn't identified during the planning activities has a chance to make to the Stakeholder Register and also have a Stakeholder Management Strategy.

As a summary of the discussion around the Initiating process and potential gaps in the poor planning, the following table presents the main underlying issues for the poor planning root cause, the corresponding framework artifact that addresses each issue, and potential gaps identified in each one that need addressing.

ROOT-CAUSE RESULT FOR POOR PROJECT PLAN-NING	INITIATING PROCESS GROUP ARTIFACT	GAPS
1 Lack of Established Project Objectives	Project Charter	Only partially addresses the potential lack of project objectives as it assumes good inputs to the Create Project Charter process, but doesn't give specifics on how to generate those inputs.
1.1 Poor Communication with Stakeholders	Stakeholder Management Strategy	Doesn't address the issue of different stakeholders' backgrounds; more evident on T&M projects.
1.2 Missing Stakeholders	Stakeholder Register	Doesn't present a methodology to avoid errors of omission and missing stakeholders due to the potential interdependencies between all technical disciplines of T&M projects.
1.3 Lack of High-Level Scope of Work	Project Charter	Project Statement of Work input to Create Project Charter might generate "garbage in, garbage out" scenario. No current methodology to minimize this risk described by current version of PMBOK®.

ROOT-CAUSE RESULT FOR POOR PROJECT PLAN- NING	INITIATING PROCESS GROUP ARTIFACT	GAPS
1.4 Lack of Accept- ance Criteria	None	All
2 Poor Risk Identifi- cation	None	All
2.1 Unbalanced Business and Tech- nical Risk Identifi- cation	None	All
2.2 Not All Appro- priate Stakeholders Involved in Risk Identification	None	All
2.3 Project Manager Tries to Do It by Her- self	None	All
2.4 Risk Register Is Not Made a Living Document	None	All
2.5 No Formal Risk Identification Proc- ess Done as Part of the Precontract Phase by Service Organizations	None	All
2.6 Lack of Under- standing of the Importance of Risk Identification	None	All

ROOT-CAUSE RESULT FOR POOR PROJECT PLANNING	INITIATING PROCESS GROUP ARTIFACT	GAPS
3 Lack of Validated Assumptions	Project Charter	Project Statement of Work input to Create Project Charter might generate "garbage in, garbage out" scenario. No current methodology to minimize this risk described by current version of PMBOK®.
4 Lack of Stakeholder Buy-In to the Plan	None	All

Table 3. PM Framework Initiating versus Poor Planning

The next step of the analysis is to dive into the Planning process group in hope of filling up the middle column of table 3, especially for the lines that no process group artifact from the Initiating process group was identified.

Planning Process Group

The PMBOK® framework has several processes dedicated to planning activities. For the sake of our discussion, we will be focusing solely on the ones that have direct correlation with the underlying issues we have identified as the main drivers for poor project planning.

In summary, all activities incorporated by this process group culminate in a single output artifact, the Project Management Plan. According to the PMBOK®, the Project Management Plan defines how the project is executed, monitored, controlled, and closed. The Project Management Plan documents the collection of outputs from the planning processes, including:

- Project management processes selected by the project management team
- Level of implementation of each selected process
- Tools and techniques descriptions selected to accomplish the selected processes
- Description of how the selected process will be used for the specific project
- Description of how work will be executed to accomplish the project objectives
- Description of how change will be monitored and controlled
- Description of how configuration management will be performed
- Description of how the performance baseline will be maintained and used
- Stakeholder communication requirements and plan
- Selected project life cycle for multiphase project
- Management reviews processes for content, extent, and timing to address open issues and pending decisions

The Project Management Plan is a formal, approved document that defines how the project is executed, monitored, and controlled. It may be summarized or detailed and may be composed of one or more subsidiary management plans and other planning documents.

Figure 13 shows all processes that make up the Planning process group. The ones of special interest to our discussion are highlighted, though all the processes presented as part of the Planning process are important and should be implemented during planning activities of T&M projects.

In referring back to the underlying issues analysis performed for the Initiating process group, the first one down the list that is not being addressed at all is the lack of acceptance criteria. In looking through the processes that are part of the Planning process group, the one that intersects with that underlying issue is the Plan Quality process.

Plan Quality is the process of identifying required quality standards, the metrics that will demonstrate compliance to those standards, and the tools and techniques that will be utilized in obtaining those metrics.

This is an area of strength of the PMBOK® framework, and therefore the issue at hand is more of training and enforcing of the proper execution in the creation of a comprehensive acceptance criteria document.

Technical project managers, most of the time, get attracted by the technical issues of the project at hand and tend to dismiss surrounding project management activities that do add immense value to the sound management of the project, acceptance criteria being one of them. Most of them do understand the value of acceptance criteria; however, very few understand this activity as being part of the initial project planning tasks.

The standard practice has been to get only very high-level acceptance criteria in place during the planning phase, to the point of actually missing some lower-level, yet fundamental criteria that only bubble to the surface once the entire implementation phase is complete and verification is about to start.

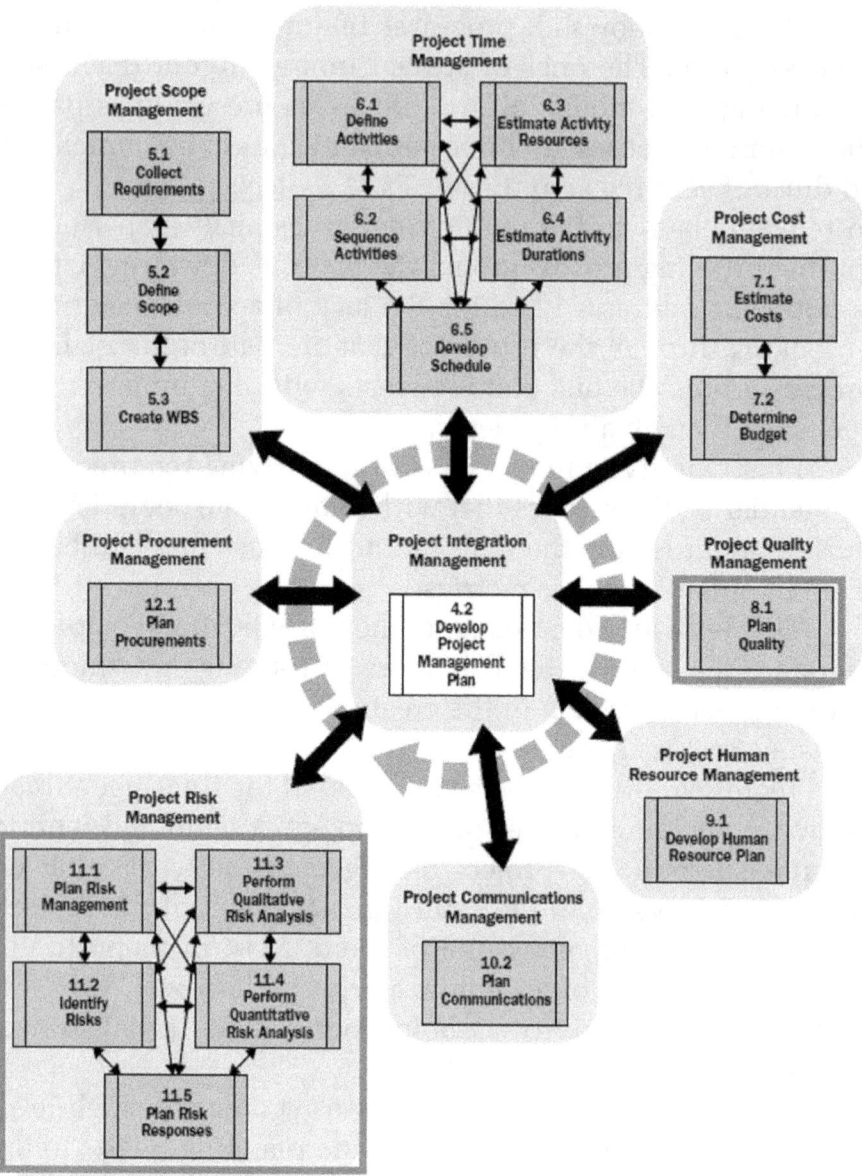

Figure 13. PMBOK® Framework Planning Process Group

As it was pointed out in a previous chapter, this might lead to a catastrophic outcome, as some of those lower-level criteria could have been potentially the driver to different design and implementation paths than the ones that were chosen by the project team.

This is also very often the case when engaging a system integrator to execute the T&M project. Integrators' proposals tend to focus more on the high-level scope of work as well as the technical assumptions that were used for the initial technical approach proposed. They usually don't pay much attention to the overall acceptance criteria that will determine project completion with the client.

Assume as an illustration a proposal that states a final site acceptance test is to be carried out in the deployed system at the client site without providing extra details of what that test is composed of. Assume now that the client has very stringent requirements for contractor presence on site. The client requires any contractor that will be on site to sit through eight hours of safety training and that the contractor can execute the work only over the weekend.

This will add significant cost to the integrator that was not captured as part of the cost to execute the site acceptance test.

A second area within the planning process group that is worth mentioning is Project Risk Management. According to the PMBOK®, Project Risk Management is the set of processes that includes risk management planning, identification, analysis, responses, and monitoring of project risks. Its objective, as listed in the PMBOK®, is to increase the probability and impact of positive events and decrease probability and impact of negative events in the project. This is very much in line with what was presented in a previous chapter around the importance of identifying and monitoring project risks. However, let's now cross-reference the identified underlying issues around poor risk identification presented at the beginning of this chapter and try to find correlation with PMBOK®'s project risk management processes. Figure 14 presents PMBOK®'s summary of its Project Risk Management processes.

Unbalanced Business and Technical Risk Identification: This can qualify as one of the so-called soft skills by the project manager. As seen in a previous chapter, this plays into the psychology of the people who make good candidates to become project managers. It was mentioned that these individuals tend to pay more attention to business risks, while technical-minded people usually gravitate toward paying more attention to technical risks. The PMBOK® framework Risk Management process doesn't really address this issue, nor can it, as it focuses on processes around project management and assumes for the most part that an individual will be the sole project manager for a project. This, however, should be considered a gap in the framework as the proposed processes and organizational structure of a project under the PMBOK® proposed framework doesn't address this very important issue.

Not All Appropriate Stakeholders Involved in Risk Identification: The PMBOK® does address this issue partially as one of the inputs to the Plan Risk Management process is the communication management plan. The communication management plan defines the interactions that will occur in the project and determines who should be available for risk analysis and planning activities. However, much like the case for the underlying issue *Missing Stakeholders* that was linked to the Initiating process group, this identification of the appropriate stakeholders is trickier on T&M projects due

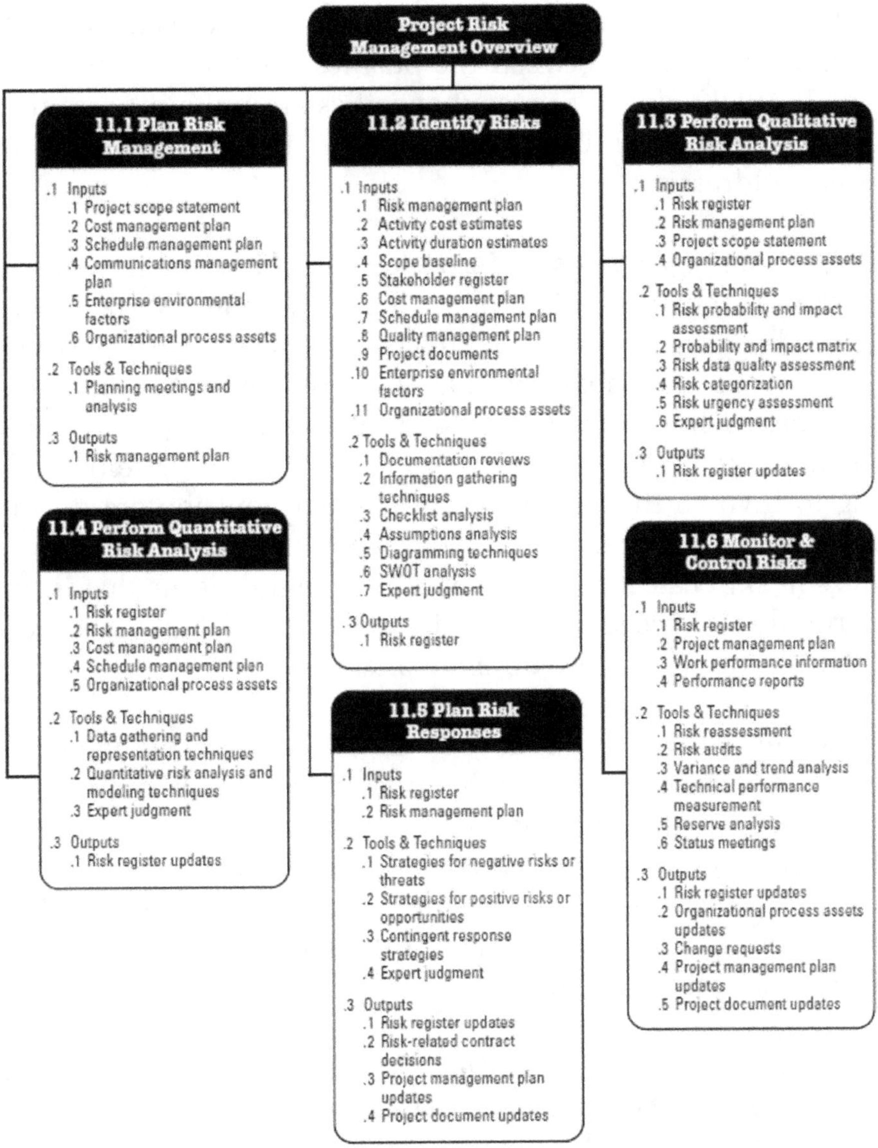

Project Risk Management Overview

11.1 Plan Risk Management

.1 Inputs
  .1 Project scope statement
  .2 Cost management plan
  .3 Schedule management plan
  .4 Communications management plan
  .5 Enterprise environmental factors
  .6 Organizational process assets

.2 Tools & Techniques
  .1 Planning meetings and analysis

.3 Outputs
  .1 Risk management plan

11.2 Identify Risks

.1 Inputs
  .1 Risk management plan
  .2 Activity cost estimates
  .3 Activity duration estimates
  .4 Scope baseline
  .5 Stakeholder register
  .6 Cost management plan
  .7 Schedule management plan
  .8 Quality management plan
  .9 Project documents
 .10 Enterprise environmental factors
 .11 Organizational process assets

.2 Tools & Techniques
  .1 Documentation reviews
  .2 Information gathering techniques
  .3 Checklist analysis
  .4 Assumptions analysis
  .5 Diagramming techniques
  .6 SWOT analysis
  .7 Expert judgment

.3 Outputs
  .1 Risk register

11.3 Perform Qualitative Risk Analysis

.1 Inputs
  .1 Risk register
  .2 Risk management plan
  .3 Project scope statement
  .4 Organizational process assets

.2 Tools & Techniques
  .1 Risk probability and impact assessment
  .2 Probability and impact matrix
  .3 Risk data quality assessment
  .4 Risk categorization
  .5 Risk urgency assessment
  .6 Expert judgment

.3 Outputs
  .1 Risk register updates

11.4 Perform Quantitative Risk Analysis

.1 Inputs
  .1 Risk register
  .2 Risk management plan
  .3 Cost management plan
  .4 Schedule management plan
  .5 Organizational process assets

.2 Tools & Techniques
  .1 Data gathering and representation techniques
  .2 Quantitative risk analysis and modeling techniques
  .3 Expert judgment

.3 Outputs
  .1 Risk register updates

11.5 Plan Risk Responses

.1 Inputs
  .1 Risk register
  .2 Risk management plan

.2 Tools & Techniques
  .1 Strategies for negative risks or threats
  .2 Strategies for positive risks or opportunities
  .3 Contingent response strategies
  .4 Expert judgment

.3 Outputs
  .1 Risk register updates
  .2 Risk-related contract decisions
  .3 Project management plan updates
  .4 Project document updates

11.6 Monitor & Control Risks

.1 Inputs
  .1 Risk register
  .2 Project management plan
  .3 Work performance information
  .4 Performance reports

.2 Tools & Techniques
  .1 Risk reassessment
  .2 Risk audits
  .3 Variance and trend analysis
  .4 Technical performance measurement
  .5 Reserve analysis
  .6 Status meetings

.3 Outputs
  .1 Risk register updates
  .2 Organizational process assets updates
  .3 Change requests
  .4 Project management plan updates
  .5 Project document updates

Figure 14. PMBOK® Project Risk Management Process

to the multidiscipline aspect of its technical subjects as well as the interaction between the business and technical areas. The PMBOK® doesn't present specifics around a process that could potentially mitigate this problem, thus being a gap that needs addressing.

Project Manager Tries to Do It by Herself: As mentioned in a previous chapter, this issue is most prominent in technical individuals who were made project managers of T&M projects. These types of professionals, as they come from the technical trenches, believe they are in the best position to perform risk identification, analysis, and response planning, therefore failing to involve the project team as other contributors during these activities. This, much like the previous item, belongs to the soft skill category; however, it is a tangible issue that is not currently addressed by the PM framework. For this reason, it is identified as a gap in the framework.

Risk Register Is Not Made a Living Document: Actually, the PMBOK® is quite strong on this topic. It does highlight the importance of constant monitoring and control of the risk register, having a separate process, called Monitor and Control Risks, devoted to this very issue. This can be classified then as a training issue in that area of the PMBOK® framework by technical project managers.

No Formal Risk Identification Process Done as Part of the Precontract Phase by Service Organizations: This topic was presented in detail by the chapter on engaging system integrators to execute the T&M project, but will be repeated here in passing due to its importance to the overall project results. Usually, a thorough risk identification is considered to be almost an afterthought by both integrator and client organizations, happening after the contract has been awarded as part of the project planning activities. An obvious issue with this is that, at that point, both organizations are already knee-deep in commitments for the project, usually too late to retreat.

Lack of Understanding of the Importance of Risk Identification: This is another one of the so-called soft skills in technical project management. This again is something that plagues most technical professionals who were made project managers. Very few technical project managers understand the importance of the risk register and risk analysis. The PMBOK® is actually very thorough on this topic; therefore, it is a training issue on these practices.

Lack of Stakeholder Buy-In to the Plan: The PMBOK® emphasizes that the main output of the Planning process group, the Project Management Plan, is a formal document that will describe how the project will be managed. As such, it does indeed call for stakeholders' signatures as a seal for the buy-in process. This is something that is not widely adopted by the T&M project management community, other than the approval of the technical proposal when integrators are engaged. One speculates the reason for this being the fact, again, that technical resources who are made project managers don't particularly buy into the discipline required by professional project management. Thus, this can be called a training issue and not necessarily a framework gap.

This concludes the gap analysis between the Planning process group and poor project planning underlying issues. Table 4 summarizes the findings.

ROOT-CAUSE RESULT FOR POOR PROJECT PLANNING	PLANNING PROCESS GROUP ARTIFACT	GAPS
1 Lack of Established Project Objectives	Quality Plan	Partially addresses this via Quality Plan and how it addressed lack of acceptance criteria
1.1 Poor Communication with Stakeholders	None	
1.2 Missing Stakeholders	None	
1.3 Lack of High-Level Scope of Work	None	
1.4 Lack of Acceptance Criteria	Quality Plan	Training issue of technical project managers
2 Poor Risk Identification	Risk Management Processes	Partially addresses this via Project Risk Management processes, though some issues are not addressed by the PM framework
2.1 Unbalanced Business and Technical Risk Identification	None	The PM framework doesn't have an answer for this issue as part of its process or organizational structure

ROOT-CAUSE RESULT FOR POOR PROJECT PLANNING	PLANNING PROCESS GROUP ARTI-FACT	GAPS
2.2 Not All Appropriate Stakeholders Involved in Risk Identification	Plan Risk Management	Partially addressed but with same gap as in the case of missing stakeholders partially addressed by the Initiating process group. No specifics around a process that could potentially mitigate this problem
2.3 Project Manager Tries to Do It by Herself	None	Soft skill that is not addressed by the PM framework
2.4 Risk Register Is Not Made a Living Document	Plan Risk Management	Training issue
2.5 No Formal Risk Identification Process Done as Part of the Pre-contract Phase by Service Organizations	None	All
2.6 Lack of Understanding of the Importance of Risk Identification	Plan Risk Management	Training issue
3 Lack of Validated Assumptions	None	All

ROOT-CAUSE RESULT FOR POOR PROJECT PLANNING	PLANNING PROCESS GROUP ARTIFACT	GAPS
4 Lack of Stakeholder Buy-In to the Plan	Project Management Plan	Training issue

Table 4. PM Framework Planning versus Poor Planning

The following table presents an overall summary of the gap analysis between the PMBOK® framework and the poor project planning driver of failed T&M projects.

ROOT-CAUSE RESULT FOR POOR PROJECT PLANNING	Addressed	Not Addressed	Partially Addressed
1 Lack of Established Project Objectives			X
1.1 Poor Communication with Stakeholders			X
1.2 Missing Stakeholders			X
1.3 Lack of High-Level Scope of Work			X
1.4 Lack of Acceptance Criteria			X
2 Poor Risk Identification			X

ROOT-CAUSE RESULT FOR POOR PROJECT PLANNING	Addressed	Not Addressed	Partially Addressed
2.1 Unbalanced Business and Technical Risk Identification		X	
2.2 Not All Appropriate Stakeholders Involved in Risk Identification			X
2.3 Project Manager Tries to Do It by Herself		X	
2.4 Risk Register Is Not Made a Living Document	X		
2.5 No Formal Risk Identification Process Done as Part of the Precontract Phase by Service Organizations		X	
2.6 Lack of Understanding of the Importance of Risk Identification	X		
3 Lack of Validated Assumptions			X
4 Lack of Stakeholder Buy-In to the Plan	X		

Table 5. Summary of PM Framework versus Poor Planning

The next step is to repeat the analysis performed for the poor planning root causes, but now focusing on the lack of established requirements root causes and the PMBOK® framework.

Lack of Established Requirements

Table 6 summarizes all the lack of established requirements root-cause items that we should attempt to trace to one or multiple existing PMBOK® processes. This list displays all the items captured by the diagram presented earlier in this chapter.

ROOT-CAUSE RESULT FOR LACK of WELL-ESTABLISHED REQUIREMENTS
1. Lack of Well-Defined Project Objectives
2. Missing Stakeholders
3. Poor Communication with Stakeholders
4. Users Don't Know What They Want
5. Errors of Omission
5.1 Lack of Solid Process on How to Gather Requirements
6. Lack of Attention to Historical Requirements
6.1 Lack of a Requirements Management System
6.2 Lack of Organization Process Assets
6.3 Lack of an Organization Level Controlling Body

Table 6. Identified Root-Cause Items for
Lack of Well-Established Requirements

The first root cause from the lack of established requirements listed in table 6 is the lack of well-defined project objectives. Incidentally, this is also a root cause for poor planning.

As seen previously, when using the poor planning lenses, the lack of well-defined project objectives will drive the planning activities to have either an incomplete or flat-out wrong set of high-level goals that will lead to an unsuccessful project.

Also, as stated in a previous chapter, the project objectives of complex T&M projects determine the high-level business and technical goals for the project. It can loosely be seen as the stakeholder's wish-list end results for the project deliverables at both business and technical levels. The project requirements can be seen as a deeper level of abstraction of the project objectives. This deeper level would define in finer details what the project deliverables must implement. Thus, it becomes natural to see the requirements as having the project objectives as their foundation. What this sentence means in practical terms is that, much like a building that needs a solid foundation, project requirements need solid project objectives to be used as starting and arrival points.

The PMBOK® framework gap analysis for lack of well-defined project objectives was already performed under the poor planning analysis and is valid for the lack of established requirements as well.

The next two issues down the list, missing stakeholders and poor communication with stakeholders, were also analyzed under the PMBOK® framework perspective for the poor planning root cause. The analysis is also valid for the lack of well-established requirements. Table 7 repeats, for convenience, the findings of this analysis in regard to those three issues for lack of well-established requirements.

ROOT-CAUSE RESULT FOR LACK OF WELL-ESTABLISHED REQUIRE-MENTS	PLANNING PROCESS GROUP ARTIFACT	GAPS
1 Lack of Established Project Objectives	Quality Plan	Partially addresses this via Quality Plan and how it addressed lack of acceptance criteria
2 Poor Communication with Stakeholders	None	All
3 Missing Stakeholders	None	All

Table 7. PM Framework Planning versus Three First Issues of Lack of Well-Established Requirements

The next issue down the list of lack of well-established requirements is the infamous *Users Don't Know What They Want.* As mentioned in a previous chapter, it is very common in T&M projects that the end users usually have only an initial idea of what the final deliverables should look like at the early stages of project planning. When the design and later the implementation phases start, users start to have a better idea about their desired outcome. Usually by that time, the project baseline of scope, budget, and schedule has already been set, and the project outcome cannot be anything other than failure.

In referring back to the PMBOK® framework Planning process group, it does call out a process named Define Scope. As defined by the PMBOK®, Define Scope is the process of developing a detailed description of the project and/or product. The inputs to this process are the project charter, the requirements documentation, and the Organizational process assets.

The first immediate issue with this definition is the assumption that the requirements documentation will be an input to the process, or, in other words, the users will know what they want and will define the project scope once the requirements documentation has been defined. And once the scope has been fully defined, the requirements will be very well established. Can you see a catch-22?

The PMBOK® mentions the following tools and techniques for use in defining the scope: expert judgment, product analysis, alternatives identification, and facilitated workshops. These are all good tools to be applied in the process of defining the project scope, so detailed definitions follow.

Expert judgment is basically the involvement of domain experts to analyze the information needed to develop the scope statement. This is especially important in T&M projects, since, as presented in previous chapters, it involves several engineering disciplines in its technical domain as well reaching into several other areas important to the client's business. The main suggestion that the PMBOK® provides in regard to this topic is for the project manager to make sure scope definition is a collaborative effort that definitely involves subject matter experts; this is definitely good advice.

Product analysis is applicable on projects that have a product as a deliverable, definitely the case in T&M projects. The PMBOK®, however, doesn't extend its explanation on what that is, other than just saying it includes techniques such as product breakdown, system analysis, requirements analysis, system engineering, value engineering, and value analysis. This is basically where the rubber meets the road in several of the issues around lack of well-established requirements for T&M projects. It is not something addressed by the PMBOK® framework, being therefore a gap to be identified in our analysis.

The next tool to be described under the Define Scope process is alternatives identification. This is also another very important activity that is not always carried out by the project team on T&M projects. The main idea is to identify alternative designs and technical approaches as possibilities to be used in the implementation of the T&M system. This is extremely valuable because the first approach identified as a possible solution is not always the best one. Only by looking at the problem through several vantage points will the best solution come to light. Remember that sometimes we use the hammer mentality, in which everything becomes a nail. Having several tools at our disposal is the best setup for the construction of the best T&M system possible.

The PMBOK® doesn't present a methodology to execute this analysis either, being therefore another gap. Table 8 captures all the gaps identified in the PMBOK® in regard to the *Users Don't Know What They Want* issue.

ROOT-CAUSE RESULT FOR LACK OF WELL-ESTABLISHED REQUIREMENTS	PLANNING PROCESS GROUP ARTIFACT	GAPS
4 Users Don't Know What They Want	Define Scope	Assumes requirements documentation is complete. Doesn't define how the product analysis and alternative identification are to be carried out

Table 8. PM Framework Planning versus
Users Don't Know What They Want Issue

The next issue to be analyzed is the lack of solid process on how to gather requirements, which is the biggest driver to errors of omission. In going back to the PMBOK® Planning process group, there is a process named Collect Requirements that is the best candidate to address this issue. As defined by the PMBOK®, Collect Requirements is the process of defining and documenting stakeholders' needs to meet the project objectives. It sounds promising. Its inputs are the project charter and the stakeholder register and it outputs the requirements documentation, the requirements management plan, and the requirements traceability matrix.

The PMBOK® also mentions that requirements are to be categorized as what they call project requirements and product requirements, whereas the project requirements include the business requirements and all requirements that surround the actual product requirements. The product requirements include the technical requirements and nonfunctional requirements for the product. This definition finds a good parallel on T&M projects, where there is a "product," the T&M system and the surrounding requirements, including the business requirements.

The PMBOK® defines the following tools and techniques for collecting requirements: interviews, focus groups, facilitated workshops, questionnaires, observations, and prototypes. These are all good activities that can and should be used to collect requirements.

However, as stressed in previous chapters, T&M systems usually have a reach that involves stakeholders from multiple backgrounds. Since it is a multidisciplinary engineering effort, it may include technical stakeholders of many different areas such as software, electrical, mechanical, chemical, RF, and material science. Also, the T&M system is usually part of an NPI effort, which touches product development and supporting departments such as quality and manufacturing. Ultimately, it influences business aspects of the client's organization, which may touch financial, accounting, planning, etc. All these stake-holders with different backgrounds will most likely have difficulty understanding what the requirements are currently describing and, potentially, in describing what their requirements are to the person collecting them, who probably has yet another background.

All of the tools and techniques mentioned by the PM-BOK® don't necessarily address the issue at hand; therefore, this is identified as a gap.

ROOT-CAUSE RESULT FOR LACK OF WELL-ESTABLISHED REQUIREMENTS	PLANNING PROC-ESS GROUP ARTI-FACT	GAPS
5.1 Lack of Solid Process on How to Gather Requirements	Collect Requirements	Doesn't define a solid process to bridge the gap between the T&M project stakeholders' different backgrounds

Table 9. PM Framework Planning versus *Lack of Solid Process on How to Gather Requirements* Issue

The last three issues shown under the lack of attention to historical requirements are lack of a requirements management system, lack of organization process assets, and lack of an organization level controlling body.

In regard to the requirements management system, the PM-BOK® has a process called Direct and Manage Project Execution, which is part of the Execution process group. This process mentions the importance of a project management system, which usually includes some sort of requirements management system.

Management of requirements is one of the most important activities to be performed within the scope of a T&M system, along with risk identification and management. As seen
in previous chapters, due to the complexity of T&M systems,
requirements have a tendency of evolving as more is learned
about the system and its challenges.

The lack of a requirements management system will make
the task of properly managing the system requirements very
difficult. Such management systems usually provide a database where all requirements, not only the ones that are active,
but also the ones that are deprioritized, can be maintained. It is
extremely important that this information is not only kept up
to date, but that it is acted upon by the project team.

Furthermore, more than just a requirements management
system is needed; also necessary is a methodology to reprioritize
requirements and make sure these changes are rolled back to the
project plan and all of its artifacts, as well as to the risk registry.
The PMBOK® framework doesn't provide a good systematic approach on how to accomplish this; therefore, it is flagged as a gap.

ROOT-CAUSE RESULT FOR LACK OF WELL-ESTABLISHED REQUIREMENTS	PLANNING PROCESS GROUP ARTIFACT	GAPS
6.1 Lack of a Requirements Management System	Direct and Manage Project Execution Process	Doesn't define a methodology to reprioritize requirements and make sure these changes are rolled back to the project plan and all of its artifacts, as well as to the risk registry

Table 10. PM Framework Planning versus Lack of a Requirements
Management System

Lack of an organization level controlling body is the next issue. As it was previously mentioned, the PMBOK® defines the concept of a PMO, or, Project Management Office, as being that body. The PMBOK® defines the PMO as an organizational body or entity assigned various responsibilities related to the centralized and coordinated management of the projects under its domain.

This, however, is not something that can be fixed at the T&M project manager's level. It is something that needs to be addressed at the organization level as a whole. PMI has a lot of good material on how to set up a PMO in an organization, therefore being a great source of knowledge in the subject.

ROOT-CAUSE RESULT FOR LACK OF WELL-ESTABLISHED REQUIREMENTS	PLANNING PROCESS GROUP ARTIFACT	GAPS
6.3 Lack of an Organization Level Controlling Body	PMO	None

Table 11. PM Framework Planning versus *Lack of an Organization Level Controlling Body* Issue

This next issue to be cross-referenced against the PMBOK® is the lack of organization process assets. The PMBOK® has good material about the organizational influences on project management and it does refer to the organizational process assets as one of those areas of influence.

It does talk about the processes and procedures and the corporate knowledge base that are part of these assets and is a good reference for organizations looking to improve on this recognized issue.

ROOT-CAUSE RESULT FOR LACK OF WELL-ESTABLISHED REQUIREMENTS	PLANNING PROCESS GROUP ARTIFACT	GAPS
6.2 Lack of Organization Process Assets	Organizational Influences on Project Management	None

Table 12. PMBOK® Framework Planning versus *Lack of Organization Process Assets* Issue

Tables 13 and 14 summarize the PMBOK® framework coverage of the identified issues related with the lack of well-established requirements.

ROOT-CAUSE RESULT FOR LACK OF WELL-ESTABLISHED REQUIREMENTS	PLANNING PROCESS GROUP ARTIFACT	GAPS
1 Lack of Established Project Objectives	Quality Plan	Partially addresses this via Quality Plan and how it addressed lack of acceptance criteria
2 Poor Communication with Stakeholders	None	All
3 Missing Stakeholders	None	All
4 Users Don't Know What They Want	Define Scope	Assumes requirements documentation is complete. Doesn't define how the product analysis and alternative identification are to be carried out

5 Errors of Omission	Collect Requirements	Doesn't define a solid process to bridge the gap between the T&M project stakeholders' different backgrounds
5.1 Lack of Solid Process on How to Gather Requirements	Collect Requirements	Doesn't define a solid process to bridge the gap between the T&M project stakeholders' different backgrounds
6 Lack of Attention to Historical Requirements		
6.1 Lack of a Requirements Management System	Direct and Manage Project Execution Process	Doesn't define a methodology to reprioritize requirements and make sure these changes are rolled back to the project plan and all of its artifacts, as well as to the risk registry
6.2 Lack of Organization Process Assets	Organizational Influences on Project Management	None
6.3 Lack of an Organization Level Controlling Body	PMO	None

Table 13. PMBOK® Framework versus
Lack of Well-Established Requirements

ROOT-CAUSE RESULT FOR LACK OF WELL-ESTABLISHED REQUIREMENTS	Addressed	Not Addressed	Partially Addressed
1 Lack of Established Project Objectives			X
2 Missing Stakeholders			X
3 Poor Communication with Stakeholders			X
4 Users Don't Know What They Want			X
5 Errors of Omission		X	
5.1 Lack of Solid Process on How to Gather Requirements		X	
6 Lack of Attention to Historical Requirements			X
6.1 Lack of a Requirements Management System			X
6.2 Lack of Organization Process Assets	X		
6.3 Lack of an Organization Level Controlling Body	X		

Table 14. Summary of PMBOK® Framework versus Lack of Well-Established Requirements

As some final comments on this section, it is important to make sure the reader understands that what is being presented here doesn't intend to hint that the PMBOK® framework is incomplete in its nature. What is being shown is that since the PMBOK® framework is intended to cover projects of many sizes and industries, as well as to keep itself bound to project management subjects only, it can and should be utilized in conjunction with a complementary set of processes when applied to management of T&M projects. The intention of this book is to show what processes and best practices can be utilized by organizations in order to raise its chances of success in executing T&M projects.

The next section presents a similar analysis for INCOSE's system engineering framework.

The Systems Engineering Framework

The second framework that will be analyzed is what we will be calling here the SE (systems engineering) framework. Much like PMI is the organization that championed project management and created the PMBOK® framework, INCOSE (International Council of Systems Engineering) is the organization responsible for the dissemination of the systems engineering concepts that were compiled into what is called the *SE Handbook*.

INCOSE defines systems engineering as an interdisciplinary approach and means to enable the realization of successful systems. As seen previously in this book, a T&M system is a highly interdisciplinary system. Therefore, chances are that the SE framework may be a good candidate for the solution we seek.

Furthermore, the *SE Handbook* states that the system engineering process has an iterative nature that supports learning and continuous improvement. As the process unfolds, system engineers uncover the real requirements and the emergent properties of the system. System complexity can lead to unexpected and unpredictable behavior of systems. Furthermore, it states that the systems engineering discipline includes both technical and management processes. This all sounds very familiar in the context of T&M systems.

One of the main tenets of the SE framework is the concept of project life cycle with decision gates. Decision gates are also often called milestones, and should answer the following questions:

- Does the project deliverable still satisfy the business case?
- Is it affordable?
- Can it be delivered when needed?

The decision gates should represent major decision points in the system life cycle and ensure that new activities are not pursued until the previously scheduled gating activities are completed.

The two main objectives for the approach of decision gates is to make sure the system being built is still in line with the business case that funded the effort, and to make sure the risk of moving forward is still within acceptable boundaries.

The *SE Handbook*'s preferred method of illustrating the concept of system life cycle is through what is known as the "Vee" (pronounced as the letter *V*) diagram.

The majority of highly qualified system integrators utilize the following Vee model as their T&M system implementation model. PDR stands for preliminary design review, CDR stands for critical design review, FAT stands for factory acceptance test, and SAT stands for site acceptance test.

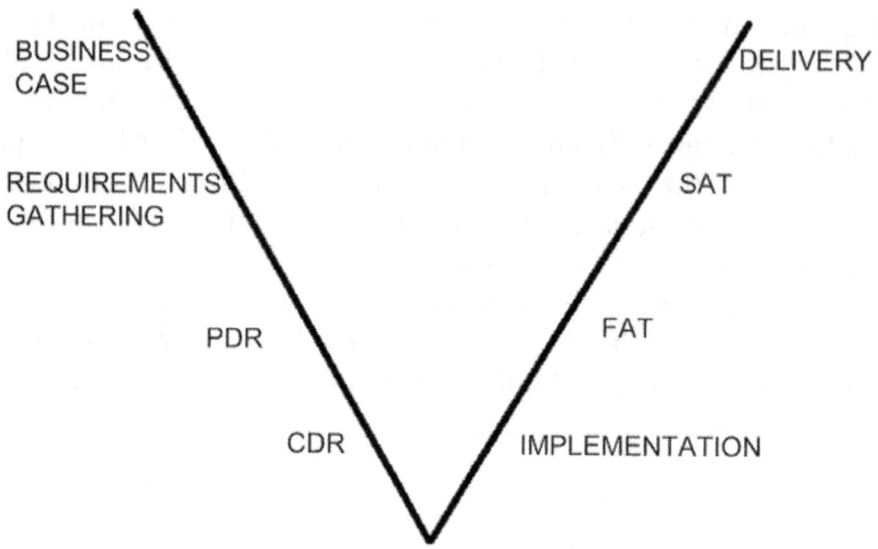

Figure 15. System Integrators Vee Diagram

The *SE Handbook* presents a set of twelve high-level processes that summarize the framework in regard to actually building the system: Stakeholder Requirements Definition, Requirements Analysis, Architectural Design, Implementation, Integration, Verification, Transition, Validation, Operation, Maintenance, and Disposal.

The framework also presents a set of what it calls the Project Processes and the Enterprise and Agreement Processes.

The Project Processes are involved with the activity of managing a system project. The following diagram shows the overlap between project management and systems engineering, as described by the *SE Handbook*.

One will notice that this framework proposes that although there is an area where tasks overlap between systems engineering and project management, there is a set of tasks that pertain to the system engineer and another set that is unique to the project manager.

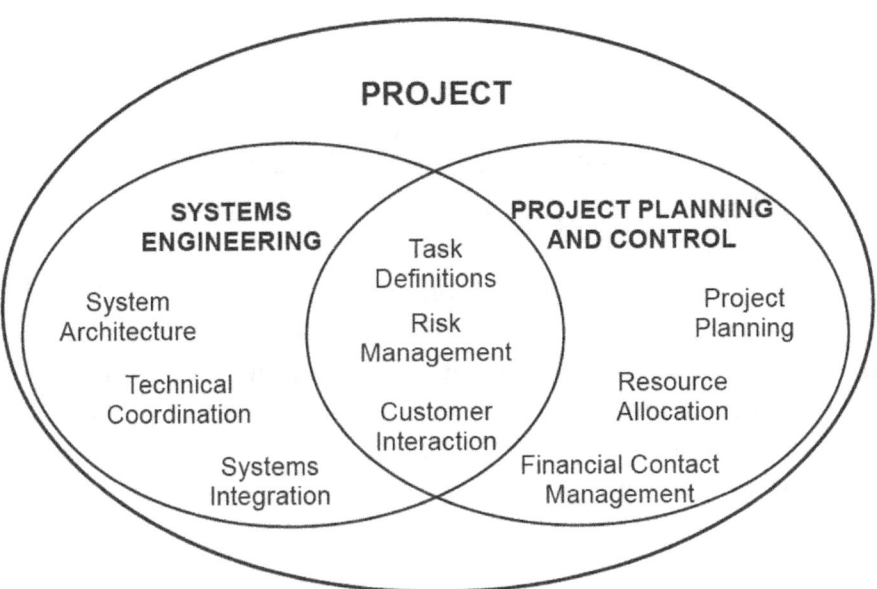

Figure 16. The Overlap Between Project Management
and Systems Engineering

This is aligned with the notion that has been presented in this book that it is extremely difficult for an organization to find a single individual who can be the systems engineer (using the SE framework terminology) who can also function as the project manager. This framework recognizes the challenge of having a single role responsible for both the technical leadership and project management aspects of a technical project.

The project processes listed in the *SE Handbook* are project planning, project assessment, project control, decision making, risk and opportunity management, configuration management, and information management.

The Enterprise and Agreement Processes basically cover the capabilities of an organization relevant to the realization of a system. It can be partially seen as a counterpart of the concepts of Enterprise Environmental Factors and Organization Process Assets as presented by the PMBOK® framework; however, it does provide some extra information on how the system is to be linked to the business value that it needs to provide to the client's organization. It can also be seen as a PMO for systems.

The processes that belong to this group are enterprise environment management process, investment management process, system life cycle processes management process, resource management process, quality management process, acquisition process, and supply process.

Figure 17 summarizes the structure of the SE framework.

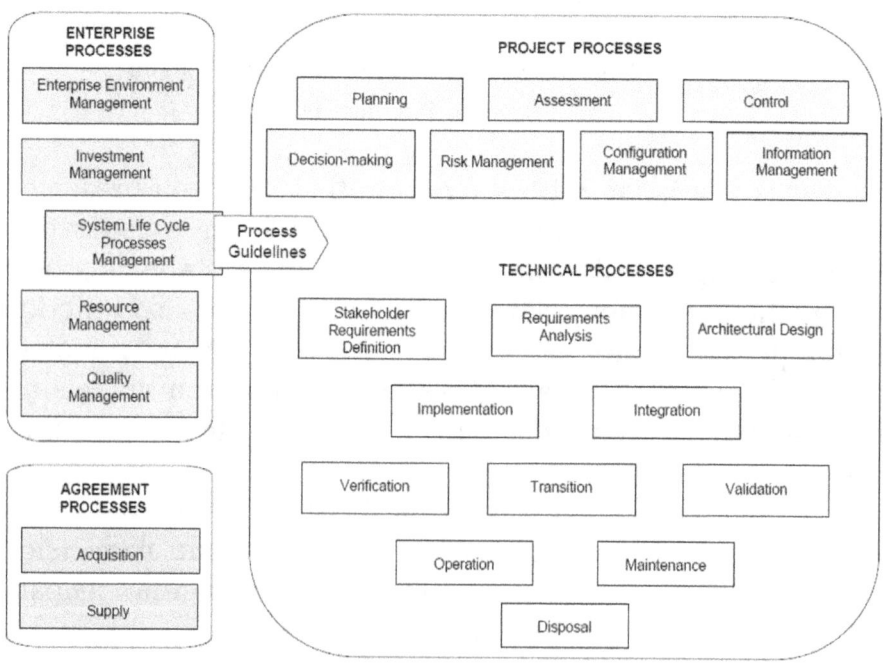

Figure 17. Summary of the SE Framework

The following subsections will explore each one of the main drivers for T&M project failure and their corresponding root causes to the relevant processes presented by the SE framework.

Poor Project Planning

As executed for the PMBOK® framework, a cross-reference analysis of each of the identified root causes for poor project planning will be performed for the SE framework.

The first high-level issue of focus for the analysis covered under this section is the lack of established project objectives. The first issue identified under this category is poor communication with stakeholders. We will perform a joint analysis with the missing stakeholders issue. The *SE Handbook* recognizes under its elicit and capture requirements section that one of the biggest challenges in this activity is the identification of the set of stakeholders from whom requirements should be elicited. As mentioned previously, this is often the case for T&M projects.

The *SE Handbook* makes mention of the same overall techniques for eliciting requirements referred by the PMBOK®. However, it goes beyond that. INCOSE has defined the concept of SysML, which stands for Systems Modeling Language.

SysML is used to model complex systems and is an extension of the family of UML (Universal Markup Language) standards that are intended to provide representations with well-defined semantics that can support model and data exchange.

SysML includes diagrams that can be used to specify system requirements, behavior, structure, and relationships. Figure 18 illustrates all available diagram types. A full description of the SysML modeling standard is beyond the scope of this book; however, it is a worthy topic of study in support of what is presented here. There is extensive literature on SysML available to the avid reader.

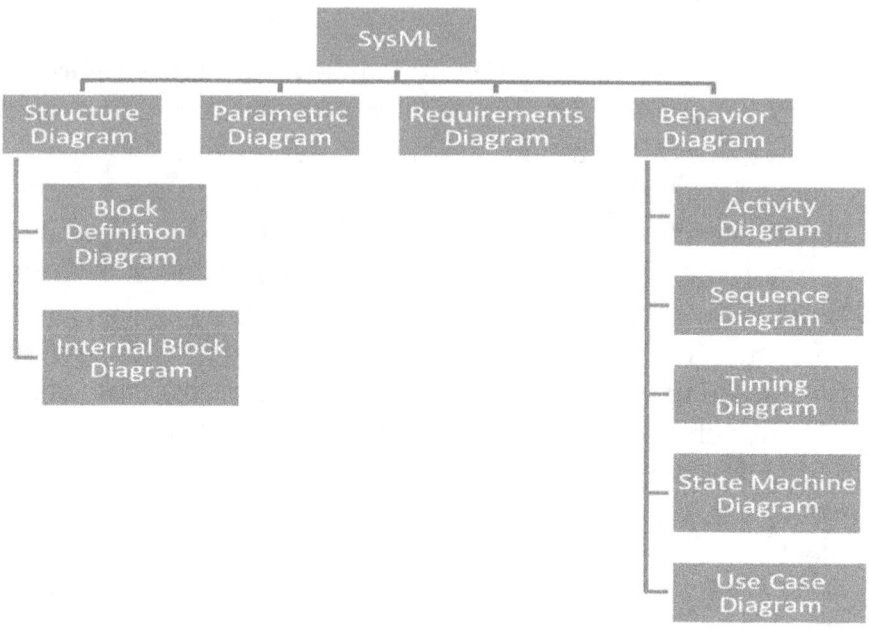

Figure 18. SysML Diagram Set

This modeling framework provides the ability for the requirements-gathering team to use many abstraction layers and look at the system to be built as well as its requirements from different vantage points. This facilitates the requirements-elicitation activity to potentially reduce its errors of omission.

Another added benefit to this approach is that it provides a visual representation of requirements, use cases, which is basically how the system will be utilized, as well as system components and interfaces with external areas. This is a great tool for communication between stakeholders of different backgrounds. It also allows for the requirements-gathering team to bring to light potential hidden external interfaces that may shed a light on stakeholders that were not initially identified.

Table 15 illustrates the coverage provided by the SE framework to both issues under analysis.

ROOT-CAUSE RESULT FOR POOR PLANNING	PLANNING PROCESS GROUP ARTIFACT	GAPS
1.1 Poor Communication with Stakeholders	SysML	None
1.2 Missing Stakeholders	SysML	None

Table 15. SE Framework Planning versus First Two Issues
of Poor Planning Root Cause

The next issue to be looked at is the lack of high-level scope of work. The *SE Handbook* includes a section named "define system capabilities and performance objectives," which at first sounds very promising.

However, the section focuses on the definition of the system architecture design activities through the definition of system constraints, such as cost and schedule, mandated use of COTS (commercial off-the-shelf) equipment, interfaces with other systems or organizations, etc.

What experience shows is that the analysis of the high-level scope of work needs to be a top-down exercise, starting from the client's high-level business value to be realized and then delving into the things that are covered by the SE framework. Therefore, this issue is being only partially covered by the SE framework.

The last issue that concludes the lack of established project objectives is lack of acceptance criteria. This is well covered by the quality management process of the SE framework. The *SE Handbook* states that the purpose of the quality management process is to make visible the goals of the enterprise toward customer satisfaction. This touches high level business goals as, the overall client's organization is the customer of a T&M system, and making sure its objectives are being fulfilled by the system would be the ultimate acceptance criteria. Figure 19 illustrates the high-level concepts behind the quality management process. One of the challenges in real-life applications, though, is to make sure that the acceptance criteria for the business value are being captured. SysML is a potentially good candidate to bring those criteria to light.

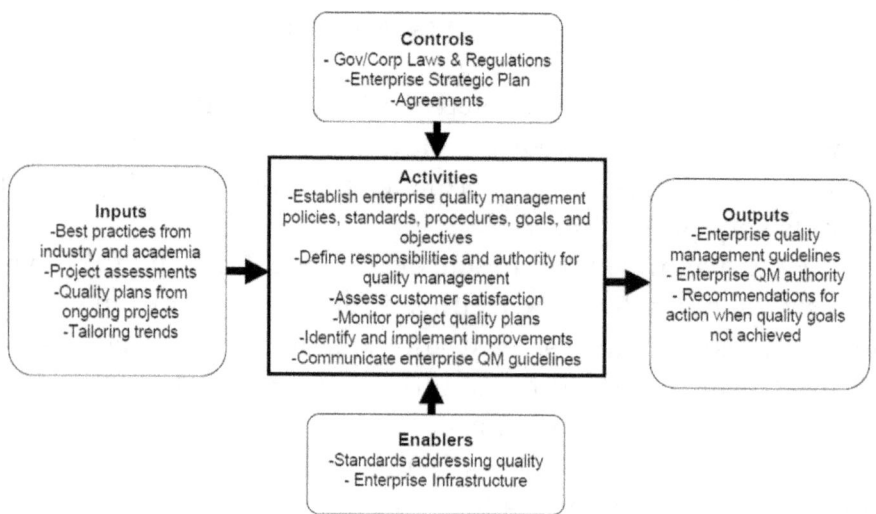

Figure 19. SE Framework Quality Management Process

Table 16 summarizes the two issues analyzed by the paragraphs above.

ROOT-CAUSE RESULT FOR POOR PLANNING	PLANNING PROCESS GROUP ARTIFACT	GAPS
1.3 Lack of High-Level Scope of Work	Define system capabilities and performance objectives section of SE Handbook	Partially covered as this section doesn't include definition of scope based on the high-level business goals
1.4 Lack of Acceptance Criteria	Quality Management Process and SysML	None

Table 16. SE Framework Planning versus First Two Issues of Poor Planning Root Cause

The next issues to be cross-referenced with the SE framework are the ones that relate to poor risk identification. The *SE Handbook* has a section named Risk and Opportunity Management that attempts to address risk. The SE framework places risk in four different categories: technical, cost, schedule, and programmatic risks. The first three are fairly self-explanatory. What the framework calls programmatic risks are basically what this book refers to the business risks, which are risks that are usually the by-product of decisions made by personnel at higher levels of authority, usually the high-level management of the company.

The first one to be looked at is the unbalanced business and technical risk identification. The SE framework has a process called risk identification; however, the process doesn't make a distinction between business and technical risks. In fact, the only mention to business risks is through the definition of the so-called process risks. The framework doesn't provide any details on how to prevent this issue from happening. It is thus flagged as a gap.

In regard to not all appropriate stakeholders being involved in risk identification, the *SE Handbook* has a section named Risk Assessment that partially addresses the issue. It does mention that expert interviews are an efficient way of identifying risks; however, it doesn't delve into details on how to do it.

In fact, as mentioned earlier, SysML is a good tool in making sure the project stakeholders are identified. This also holds true for the stakeholders needed for risk assessment. However, the framework doesn't provide details on how to actually perform this activity.

The next issue to be looked at is *project manager tries to do it all by herself*. The fact that the SE framework recognizes that there is a clear split of responsibilities between the systems engineer and the project manager is a good setup to prevent that a single person executes risk assessment.

The next issue is the fact that the risk register is not made a living document. Strangely enough, there is no mention of this fact as part of the *SE Handbook*, and it is therefore a full gap in the framework.

The next issue relates more to system integrators' organizations. The next chapter will be devoted to cross-referencing the existing frameworks with the challenges presented in a previous section on engaging integrators in T&M projects. However, for completeness, let's address that one issue here.

It was mentioned earlier in this chapter that high-quality system integrators use a system engineering Vee diagram as the preferred model of T&M system implementation. However, these companies usually utilize a partial SE framework, which will be detailed in the next chapter. Also, not many integrators usually have the rigor to perform a formal risk identification process prior to contract award.

As mentioned in a previous chapter, all the precontract tasks that are worked on by the integrator are basically at-risk work. If they are not awarded the contract, all that work is basically a loss to the company, as its business is to sell its staff's time. As such, they always attempt to minimize the time investment up front, and usually go only as deep as it allows them to provide a fixed-price proposal to their clients. This abbreviation of effort usually comes at a price of reduced risk identification as well.

The last issue that pertains to the risk root cause is lack of understanding of the importance of risk identification. The SE framework suggests that the management body of a system project, which in our case can be called a T&M project, is to be split into two roles: the systems engineer and the project manager.

Also, the framework raises the awareness of the need for risk management in general. Adding that to the fact that now there are two high-level resources performing the management of the T&M project, raising the awareness of these two resources becomes an activity of training and enforcing the framework.

Table 17 summarizes the analysis of all issues related to the risk root cause for poor planning in regard to the SE framework.

ROOT-CAUSE RESULT FOR POOR PLANNING	PLANNING PROCESS GROUP ARTI-FACT	GAPS
2.1 Unbalanced Business and Technical Risk Identification	Risk Identification Process	All as it doesn't offer a way to balance the business and technical risks
2.2 Not All Appropriate Stakeholders Involved in Risk Identification	Risk Assessment and SysML	Partial as framework doesn't provide details on how to do it
2.3 Project Manager Tries to Do It by Herself	Split into two roles: systems engineer and project manager	None

ROOT-CAUSE RESULT FOR POOR PLANNING	PLANNING PROCESS GROUP ARTIFACT	GAPS
2.4 Risk Register Is Not Made a Living Document	No mention of risk being constantly monitored	All
2.5 No Formal Risk Identification Process Done as Part of the Precontract Phase by Service Organizations	NA	NA
2.6 Lack of Understanding of the Importance of Risk Identification	Split into two roles improve the awareness of risk identification	Training issue

Table 17. SE Framework Planning versus the Risk Issues of Poor Planning Root Cause

The next issue for the poor project planning root cause is lack of validated assumptions. Since project planning is a task that is carried out in very early stages of the project life cycle, general stakeholders and project team still didn't have a chance to learn much about the project in general. Usually, as the project is carried out, the overall knowledge about it increases, leading to a situation where more accurate forecasts and decisions can be made. At very early stages, accurate predictions and learned facts that come with the effort put into project execution are traded off by assumptions. The lack of validated assumptions happens when the assumptions made at project initiation are not validated with the project stakeholders.

Unfortunately, the SE framework doesn't make any mention to these assumptions. It assumes that the requirements elicitation involves all stakeholders and happens flawlessly in a manner to capture every detail the system needs to implement. Building of T&M systems in real life happens a little differently, though. There will always be assumptions needed in order to move things along. The expectation that all requirements will be worked out up front and that no assumptions will be needed is unrealistic. This is something that the agile model focuses on fixing, as will be seen in a later chapter. However, the SE framework does have this as a gap in its model.

ROOT-CAUSE RESULT FOR POOR PLANNING	PLANNING PROCESS GROUP ARTIFACT	GAPS
3 Lack of Validated Assumption	None	All

Table 18. SE Framework Planning versus
the *Lack of Validated Assumption* Issue

The last issue under the poor project planning root cause is the lack of stakeholder buy-in to the plan. As seen for the PMBOK® framework analysis, PMBOK® states that the main output of the planning process group, the Project Management Plan, is a formal document that will describe how the project will be managed.

The *SE Handbook* also has a project planning process. As described in the *SE Handbook*, the project planning starts with a statement of need, which is basically the project objectives counterpart in the PMBOK® framework, which is often expressed in the project proposal. The generated outputs from this process would be the work breakdown structure, project milestones, task descriptions and their corresponding completion criteria, a project budget, a project quality plan, change control methodology, risk assessment, and documentation to be produced by the project.

As one can see, this process from the *SE Handbook* is almost the equivalent of the entire planning process group as described in detail by the PMBOK®. The *SE Handbook* doesn't present nearly enough details for the planning activity to be implemented for a T&M project. Furthermore, it doesn't even mention the need for a formal buy-in from the project stakeholder.

ROOT-CAUSE RESULT FOR POOR PLANNING	PLANNING PROCESS GROUP ARTIFACT	GAPS
4 Lack of Stakeholder Buy-In to the Plan	None	All

Table 19. SE Framework Planning versus the *Lack of Stakeholder Buy-In to the Plan* Issue

In summary, the SE framework mentions project planning on very high-level terms. It almost assumes that this activity is not to be part of the framework. Conversely, the PMBOK® framework pays much closer attention to the overall activity of planning a project, in despite of its shortcomings that were identified in the previous section. As such, as will be seen in the chapter presenting the TMPM framework, the answer for the poor planning root cause will most likely be based off of the PMBOK® framework, sprinkled with specifics from the SE framework that help the areas of weakness in the PMBOK® framework, as well as some new proposed additions to the framework.

Tables 20 and 21 summarize the SE framework gap analysis for the poor planning root cause.

ROOT-CAUSE RESULT FOR POOR PLANNING	PLANNING PROCESS GROUP ARTIFACT	GAPS
1.1 Poor Communication with Stakeholders	SysML	None
1.2 Missing Stakeholders	SysML	None
1.3 Lack of High-Level Scope of Work	Define system capabilities and performance objectives section of SE Handbook	Partially covered as this section doesn't include definition of scope based on the high-level business goals
1.4 Lack of Acceptance Criteria	Quality Management Process and SysML	None
2.1 Unbalanced Business and Technical Risk Identification	Risk Identification Process	All as it doesn't offer a way to balance the business and technical risks
2.2 Not All Appropriate Stakeholders Involved in Risk Identification	Risk Assessment and SysML	Partial as framework doesn't provide details on how to do it

ROOT-CAUSE RESULT FOR POOR PLANNING	PLANNING PROCESS GROUP ARTIFACT	GAPS
2.3 Project Manager Tries to Do It by Herself	Split into two roles: systems engineer and project manager	None
2.4 Risk Register Is Not Made a Living Document	No mention of risk being constantly monitored	All
2.5 No Formal Risk Identification Process Done as Part of the Precontract Phase by Service Organizations	NA	NA
2.6 Lack of Understanding of the Importance of Risk Identification	Split into two roles improve the awareness of risk identification	Training issue
3 Lack of Validated Assumption	None	All
4 Lack of Stakeholder Buy-In to the Plan	None	All

Table 20. SE Framework versus Poor Planning Root Cause

ROOT-CAUSE RESULT FOR POOR PROJECT PLANNING	Addressed	Not Addressed	Partially Addressed
1 Lack of Established Project Objectives			X

ROOT-CAUSE RESULT FOR POOR PROJECT PLANNING	Addressed	Not Addressed	Partially Addressed
1.1 Poor Communication with Stakeholders	X		
1.2 Missing Stakeholders	X		
1.3 Lack of High-Level Scope of Work			X
1.4 Lack of Acceptance Criteria	X		
2 Poor Risk Identification			X
2.1 Unbalanced Business and Technical Risk Identification		X	
2.2 Not All Appropriate Stakeholders Involved in Risk Identification			X
2.3 Project Manager Tries to Do It by Herself	X		
2.4 Risk Register Is Not Made a Living Document		X	
2.5 No Formal Risk Identification Process Done as Part of the Precontract Phase by Service Organizations	NA	NA	NA

ROOT-CAUSE RESULT FOR POOR PROJECT PLANNING	Addressed	Not Addressed	Partially Addressed
2.6 Lack of Understanding of the Importance of Risk Identification	X		
3 Lack of Validated Assumptions		X	
4 Lack of Stakeholder Buy-In to the Plan		X	

Table 21. Summary of SE Framework versus Poor Planning

Lack of Well-Established Requirements

The last step that is missing in our root-cause analysis is the cross-referencing of each of the identified root causes for lack of well-established requirements against the SE framework. This section will focus on this analysis.

The first issue identified under lack of well-established requirements is the lack of well-defined project objectives. Unfortunately, the *SE Handbook* makes only a brief mention about the definition of the projects objectives, under its project planning process. As mentioned previously, the SE framework project planning process is not very strong, only mentioning the activities that need to take place, but without a comprehensive description on how to implement them. As such, it will be identified as a gap in the framework.

ROOT-CAUSE RESULT FOR POOR PLANNING	PLANNING PROCESS GROUP ARTIFACT	GAPS
1 Lack of Well-Defined Project Objectives	None	All

Table 22. SE Framework Planning versus Lack
of Well-Defined Project Objectives

As seen in the case of the analysis performed for the PM-BOK® framework, poor communication with stakeholders and missing stakeholders are also issues that were identified as root causes for poor planning. That analysis, when performed for the SE framework, showed that the proper utilization of SysML to model the system requirements bridges the gaps for these two issues. The summary of that analysis is repeated here for convenience.

ROOT-CAUSE RESULT FOR POOR PLANNING	PLANNING PROCESS GROUP ARTIFACT	GAPS
2 Missing Stakeholders	SysML	None
3 Poor Communication with Stakeholders	SysML	None

Table 23. SE Framework Planning versus Missing Stakeholders and Poor
Communication with Stakeholders

The next issue identified under this root cause is *users don't know what they want*. As it was mentioned earlier in this chapter, one of the challenges in T&M projects that drives the perception that users don't know what they want is the fact that it is sometimes difficult for the stakeholders to visualize the system features and interfaces early in the project life cycle.

It was also mentioned that the SysML-based require-ments-modeling approach to requirements elicitation is a great help in communicating the system requirements to the project stakeholders. The argument about improvement of the communication with stakeholders stems from the fact that when executing proper system requirements modeling, the stakeholders will be able to better visualize what the system is proposed to implement. This is exactly in line with the idea that users don't know what they want early in the project life cycle because they can't visualize well what the system will do and how the specific stakeholders' requirements are being fulfilled by it.

Therefore, even though the SE framework doesn't neces-sarily focus on addressing this issue as part of the *SE Hand-book*, the utilization of the SysML-based requirements-mode-ling technique partially bridges this gap.

ROOT-CAUSE RESULT FOR POOR PLANNING	PLANNING PROC-ESS GROUP ARTI-FACT	GAPS
4 Users Don't Know What They Want	SysML	Partially via SysML

Table 24. SE Framework Planning versus Users
Don't Know What They Want

The next issue down the list is the infamous errors of omission. The argument around errors of omission that was made in a previous chapter identified the lack of a solid proc-ess on how to gather requirements as one of the biggest drivers of errors of omission.

We are again to refer back to the SysML requirements-modeling process as the solid process that the PMBOK® framework lacks for requirements elicitation. We will mention here that SysML partially addresses the issue, since, in reality, there isn't a silver bullet against errors of omission.

Although having a solid requirements-elicitation process certainly helps the cause immensely, at the end of the day, requirements gathering is as much of an art as it is an analytical process. Even when the technique is applied, the end results are only as good as the people applying it. Especially on T&M projects and their interdisciplinary nature, the requirements engineer needs to not only understand how to best apply SysML to the case at hand and understand in depth how T&M systems are built, but, most importantly, she must have the humility to recognize when help is needed on specific domain areas and/or business issues. As in the case of any engineering process, the process is only as good as its implementation.

ROOT-CAUSE RESULT FOR POOR PLANNING	PLANNING PROCESS GROUP ARTIFACT	GAPS
5 Errors of Omission	SysML	Partially
5.1 Lack of Solid Process on How to Gather Requirements	SysML	Partially

Table 25. SE Framework Planning versus Errors of Omission

The last three issues that are shown under the lack of attention to historical requirements are lack of a requirements management system, lack of organization process assets, and lack of an organization level controlling body.

The *SE Handbook* has a full section named information management as well as a process of the same name that partially touches the requirements management system. Though these sections don't provide comprehensive information on the topic, it is clear, though, that the framework makes a valid effort to educate its followers on the need for a good information management system, which contains a requirements management system.

There is plenty of available literature on the topic and requirements management tools in the marketplace. However, as was mentioned during the analysis for the PMBOK® framework, more than just a requirements management system is needed; also needed is a methodology to reprioritize requirements and make sure these changes are rolled back to the project plan and all of its artifacts, as well as to the risk registry. Much like the case with the PMBOK® framework, the SE framework doesn't provide a good systematic approach on how to accomplish this; therefore, it is flagged as a gap.

ROOT-CAUSE RESULT FOR LACK OF WELL-ESTAB-LISHED REQUIREMENTS	PLANNING PROCESS GROUP ARTIFACT	GAPS
6.1 Lack of a Requirements Management System	Information Management Process	Doesn't define a methodology to reprioritize requirements and make sure these changes are rolled back to the project plan and all of its artifacts, as well as to the risk registry

Table 26. SE Framework versus Lack of a
Requirements Management System

The next issue to be cross-referenced against the SE framework is the lack of organization process assets.

The *SE Handbook* has a process named Enterprise Management Process to establish and maintain a set of policies and procedures to support the organization to acquire and supply products and services. This is the counterpart of PMBOK®'s organization process assets. The main idea proposed by the SE framework is for the organization to have a set of processes, procedures, and templates to establish, communicate, and continuously improve the system life cycle. This goes exactly in line with the lack of organization process assets issue at hand.

Figure 20 summarizes the process.

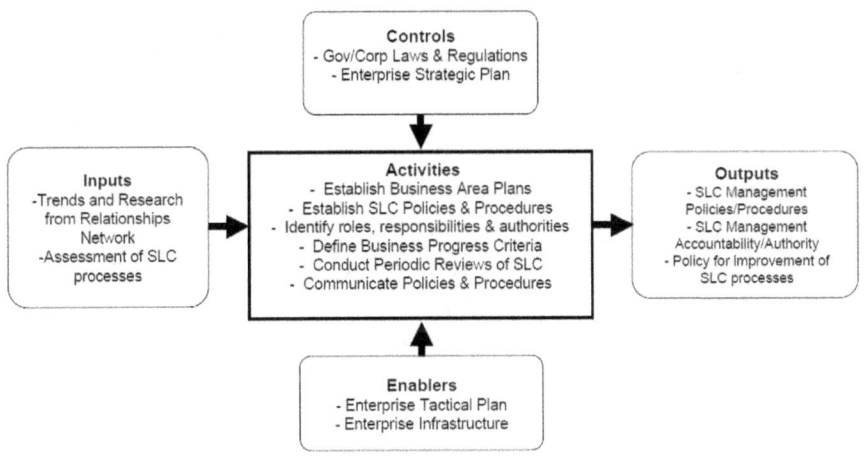

Figure 20. Enterprise Management Process: SE Framework

The *SE Handbook* does a good job at defining what the activities described above as the central point of the process should be. Per the *SE Handbook*, the establishment of business area plans is the activity to align organizational level strategic objectives to identify candidate projects that will fulfill them.

As it was seen along the course of this chapter, this is definitely something that organizations don't necessarily do enough of. This is somewhat intertwined with the lack of well-established project objectives issue that was previously described.

The *SE Handbook* also defines the other activities captured as part of the activities box above. The one of particular interest for this discussion is the establishment of system life cycle (SLC) policies, processes, and procedures. The main idea behind this concept is to make sure the SLC is consistent with the organizational business area plans as described in the previous paragraph. The overall concept is that not only is an organization best served when there is a standard system life cycle that is adopted throughout the enterprise when it comes to building T&M systems, but it should also make sure that this standard SLC is aligned with the business area plans.

An organization that is diligent in establishing such standardization benefits tremendously in all areas that surround the activities of building T&M systems. Not only is it best served in maximizing the business value for each capital investment in a T&M system, but it also sees materials cost savings and efficiency gains on engineering activities.

Assume a company that, for instance, standardizes its high-level software architecture across all T&M systems. Every new system to be built will have a fairly good basis to start from, as opposed to starting from a blank software page. This brings all sorts of advantages such as a quicker learning curve for new resources, less verification and validation efforts as now regression testing can be applied, less overall software defects as a strong and proven code base is used as the framework, a better framework for code reuse, etc.

Looking at this now under the hardware lens, assume a company standardizes in equipment from a single hardware vendor, to the extent that it is possible. This organization will have more purchasing power with that vendor, which can lead to volume discounts. It will also have to purchase fewer spare parts for the overall systems, as they will have the potential to be valid across multiple systems. It will also create an environment where the engineers will potentially have a smaller set of instruments to become experts on. Ultimately, it has the potential to establish a close partnership between the organization and the hardware vendor, which is always advantageous to the client, especially in a time of emergency.

Although the SE framework does a good job explaining the value of this, as does the PMBOK® framework, this mentality is still being developed within organizations. It is not common to see organizations that are mature in their process assets have a standardized SLC to the point of having standard frameworks that can be used across the entire enterprise when it comes to T&M systems. This is certainly a training issue not only to be targeted to test departments and T&M project managers, but should also be addressed to C-level individuals: COOs, CEOs, CTOs, etc. These individuals usually are in a better position to make organization level decisions around standardizations.

ROOT-CAUSE RESULT FOR LACK OF WELL-ESTABLISHED REQUIRE-MENTS	PLANNING PROCESS GROUP ARTI-FACT	GAPS
6.2 Lack of Organization Process Assets	Enterprise Management Process	Partially, as it requires deeper training not only for people directly involved in T&M but higher at the organization decision-making level

Table 27. SE Framework versus *Lack of Organization Process Assets* Issue

The issue that will close the analysis for lack of well-established requirements is lack of an organization level controlling body. As it was seen through the analysis of lack of organization process assets, the SE framework recommends an organizational level establishment of business area plans as well as the implementation of SLC policies, processes, and procedures that are valid throughout the organization. This strongly suggests that it can only be achieved through some sort of centralized controlling body.

Strangely enough, the SE framework doesn't suggest such centralized body, as does the PMBOK® framework with the PMO. This is then a gap in the framework; however, as it will be seen in a later chapter, merging the recommendations of both frameworks is probably good enough to bridge the gap of this issue.

ROOT-CAUSE RESULT FOR LACK OF WELL-ESTABLISHED REQUIREMENTS	PLANNING PROCESS GROUP ARTIFACT	GAPS
6.3 Lack of an Organization Level Controlling Body	None	All

Table 28. SE Framework versus Lack of an Organization Level Controlling Body Issue

Tables 29 and 30 provide the summary of findings for the SE framework analysis against the lack of well-established requirements root cause.

ROOT-CAUSE RESULT FOR POOR PLANNING	PLANNING PROCESS GROUP ARTI-FACT	GAPS
1 Lack of Well-Defined Project Objectives	None	All
2 Missing Stakehold-ers	SysML	None
3 Poor Communica-tion with Stakeholders	SysML	None
4 Users Don't Know What They Want	SysML	Partially via SysML
5 Errors of Omission	SysML	Partially via SysML
5.1 Lack of Solid Proc-ess on How to Gather Requirements	SysML	Partially via SysML
6 Lack of Attention to Historical Require-ments		
6.1 Lack of a Require-ments Management System	Information Management Process	Partially, as it doesn't define a methodology to reprioritize requirements and make sure these changes are rolled back to the project plan and all of its artifacts, as well as to the risk registry
6.2 Lack of Organiza-tion Process Assets	Enterprise Management Process	Partially, as it requires deeper training not only for people directly involved in T&M but higher at the organization decision-making level

ROOT-CAUSE RESULT FOR POOR PLANNING	PLANNING PROCESS GROUP ARTIFACT	GAPS
6.3 Lack of an Organization Level Controlling Body	None	All

Table 29. SE Framework Against the Lack of Well-Established Requirements Root Cause

ROOT-CAUSE RESULT FOR LACK OF WELL-ESTABLISHED REQUIREMENTS	Addressed	Not Addressed	Partially Addressed
1 Lack of Established Project Objectives		X	
2 Missing Stakeholders	X		
3 Poor Communication with Stakeholders	X		
4 Users Don't Know What They Want			X
5 Errors of Omission		X	
5.1 Lack of Solid Process on How to Gather Requirements		X	
6 Lack of Attention to Historical Requirements			X

6.1 Lack of a Requirements Management System			X
6.2 Lack of Organization Process Assets			X
6.3 Lack of an Organization Level Controlling Body		X	

Table 30. Summary of SE Framework versus
Lack of Well-Established Requirements

One fact that immediately jumps out at the attentive reader is the importance of SysML in addressing several gaps that were left open by the PMBOK® framework. Requirements modeling is, as will be seen in a later chapter, one of the pillars of the new TMPM framework.

The next chapter will again perform a gap analysis of the two main frameworks presented in this chapter, but now referring to the identified challenges in working with system integrators. The goal for the chapter will be to identify whether there is a framework that addresses the issues that were identified in working with integrators, or if a new framework needs to be put in place to raise the odds for a successful engagement of that type of organization.

Once that gap analysis is performed, we will have a full picture of the strengths and weaknesses for both frameworks on all the issues that were raised in the management and implementation of T&M systems. This will set the stage for the introduction of the TMPM framework, which, as will be seen, is a hybrid framework specifically tailored for T&M projects.

CHAPTER 4:
Frameworks Gaps When Engaging System Integrators

The last chapter presented the two main frameworks commonly used in managing and executing T&M projects, PMI's PMBOK® and INCOSE's SE frameworks. These frameworks were analyzed specifically around the issues that were presented in chapter 1 as the drivers for the two main root causes for T&M project failure. A complete gap analysis was presented on how each framework addressed, or attempted to address, each one of the T&M project issues presented.

Also, a previous chapter was devoted to some known challenges in engaging system integrator companies to perform T&M projects for clients. Since this engagement also affects the end result of T&M projects, and as there are hundreds of such companies in operation in today's market, it is this author's understanding that these challenges need also to be addressed, as if they themselves were a root cause for T&M project failure. The focal point for this chapter is an analysis as to whether or not the appropriate application of either or both the PMBOK® and SE frameworks could mitigate the five main challenges that were identified when a client engages an integrator company.

This gap analysis, in addition to the one presented in the last chapter of section 1, will serve as basis for section 2 of this book, which presents the proposed TMPM framework, an improved project management framework specifically tailored for T&M projects.

Client Trusts Initial Requirements Definition to Integrators

The first scenario that was identified by the last chapter as a challenge in engaging system integrators in the execution of T&M system happen when clients trust the initial project requirements definition to integrators.

The issues presented under this scenario have two root causes as presented in the last chapter: 1) not enough time spent by the integrator on the requirements-gathering activity, and 2) a potential lack of a full picture on the integrator's part in regard to the client's target business value for the T&M system to be built.

The first step of the analysis will cross-reference the two root causes above with the lack of well-established requirements root cause and gap analysis that was performed for the PMBOK® and SE frameworks. The root causes for lack of well-established requirements are presented in table 31 for convenience.

ROOT-CAUSE RESULT FOR LACK of WELL-ESTABLISHED REQUIREMENTS
1. Lack of Well-Defined Project Objectives
2. Missing Stakeholders
3. Poor Communication with Stakeholders
4. Users Don't Know What They Want
5. Errors of Omission
5.1   Lack of Solid Process on How to Gather Requirements
6. Lack of Attention to Historical Requirements
6.1 Lack of a Requirements Management System
6.2 Lack of Organization Process Assets
6.3 Lack of an Organization Level Controlling Body

Table 31. Identified Root-Cause Items for
Lack of Well-Established Requirements

What is being proposed in this text is that the two main issues of this engagement model—not enough time spent by the integrator on requirements analysis and lack of full picture on the integrator's part in regard to the client's target business value for the T&M system—are related to items one through five in table 31.

The abbreviated time investment in requirements elicitation usually means the integrator doesn't have time to unfold all potential stakeholders that should be involved in the requirements-gathering process. Stakeholder identification is usually an interactive process in a sense where the need to involve people from other roles of the client's organization becomes incrementally clearer, as the overall project requirements take shape.

Let's take an anecdotal but illustrative example. Imagine a project where the integrator identifies that the test system software will have to interface with a database as one of its requirements. This usually means that there is more work that needs to be done by the integrator on the data storage software modules than if the test data were to be dumped into a simple .csv file, for instance.

Experienced integrators have a pretty good feel for how much time that usually takes, and since they are under an abbreviated timetable to collect the project requirements, the requirements gatherer will probably stop its inquiries about data storage at that point, once the need for a database interface has been identified.

Now imagine that the database schema hasn't been defined at that point. The database schema is basically the structure of how the data will be stored in the database. Obviously, knowing the structure that the data will take in the database is a fairly important piece of information to better understand the data storage modules of the test system to be built. Not knowing that beforehand usually means there is a lot of risk at integration time for the integrator. Since a risk exists, we saw that it is good practice to keep the focus on it until the risk can be somehow mitigated. This is good project management. In this case, the way to know more about this soon-to-be-designed database would be to involve the person on the client's side who will be responsible for creating the schema. This stakeholder will most likely be missing from the stakeholder registry until it is integration time.

Postmortem analysis of hundreds of projects over fifteen years executing T&M projects shows that there is a fairly important ratio, herein defined as the risk ratio, that has a direct correlation to project failure: the number of unknown risks divided by the number of known risks. Let's further define what each one of these terms mean.

The unknown risks are those that become actual events in the course of the project that are total surprises to the project team. For instance, in a given custom hardware project for a signal conditioning board to interface with the system instrumentation, the hardware designer chose an electronic component with specified signal-to-noise ratio that was marginally close to the overall signal conditioning board's signal-to-noise ratio specification. The project team hadn't captured that as part of the project's risk registry. At integration time, the test system didn't pass its signal-to-noise ratio specification. The unknown risk of signal-to-noise ratio became an actual event at integration time.

Now, take the same custom hardware project where the hardware designer captured the signal-to-noise ratio risk at design time. This risk was formalized as part of the project's risk registry, and a contingency/mitigation plan was put in place by the project team in case the risk materialized. The mitigation plan consisted of having an unpopulated portion of the printed circuit board for an alternate component that had a lower signal-to-noise ratio but wasn't as good of a fit for the application as the original one selected. The project manager also involved the end client and made her aware of this risk. The client agreed to relax some other system specifications that were not as critical to the project as the signal-to-noise ratio was, as part of the contingency plan. At integration time, the known risk of signal-to-noise ratio materialized and the team put in place the contingency plan to address it, which was to populate the circuit board with the alternate chip and change the other specifications that were relaxed by the client at design time.

It is easy to see how the first scenario most likely caused cost and schedule overruns since the hardware designer would need to start her research to identify an alternate component, the client would need to be involved, and, potentially, the printed circuit board would have to be redesigned and respun.

Now, extrapolate this situation to large T&M projects where several unknown risks exist. These projects that show a high-risk ratio will most likely be failed projects.

Going back to our anecdotal database example, it is somewhat easy to see how a situation like this can also generate errors of omission during the requirements gathering. Maybe the database schema included a couple of table columns with data points that are not currently included as part of the test system requirements but that are important for the quality department to perform SPC (statistical process control) analysis with. These two data types would require an extra instrument to be included as part of the instrumentation requirements.

At integration time, the database engineer will be expecting those two columns to be filled in with data; however, the test system didn't even include the instrument that was needed for that type of measurement. Worse yet, the PXI chassis that housed all instrumentation didn't have any spare slots.

Now the integrator and client are faced with the extra cost of buying another PXI chassis, buying the extra instrument, developing the software routines needed for the measurement to be taken, and integrating them into the database modules. Can you spell cost and schedule overruns?

The next point worth making is that the less time invested in the process of interviewing stakeholders and going through a systematic methodology to collect requirements, the higher will be the chances that the integrator will come across the "users don't know what they want" syndrome.

As mentioned in a previous chapter, it is very common in T&M projects that the end users usually have only an initial idea of what the final deliverables should look like in the early stages of the project planning. As the design and later the implementation phases start, users start to have a better idea about their desired outcome. Usually, by that time, the project baselines of scope, budget, and schedule have been already set, and the project outcome cannot be anything else other than failure.

A more thorough requirements-elicitation process has the power to reduce that effect. The more effort that is invested in making sure the project stakeholders fully understand what the collected requirements mean to the system and how the system would be connected to the client business, the more they can actually relate to what they think the system ought to do. As will be seen in a later chapter, requirements modeling is a technique that helps tremendously in addressing this issue. Furthermore, the requirements-modeling activity is a great way to mitigate poor communication with stakeholders. The usual method of collecting requirements via creating a text document is not conducive of a good exchange between the requirements gatherer and the client's stakeholders. This will be explored in detail in a later chapter.

Lack of a full picture on the integrator's part in regard to the client's target business value for the T&M system is the second underlying issue behind this engagement model. As was mentioned in previous chapters, project objectives are the drivers of good requirements. It is useful to think of the project requirements as a deeper level of abstraction in relation to the project objectives. This means that project requirements are a more detailed view of the high-level project objectives. If the project objectives haven't been fully captured, the system requirements will definitely turn out incomplete. Garbage in, garbage out.

If the requirements gatherer doesn't fully grasp the client's target business value for the T&M system, and since the integrator was put in charge under this engagement model to collect the requirements, most likely some project objectives will not be captured. In turn, the requirements will reflect those gaps.

Tables 32 and 33 show the gap analysis that was performed in a previous chapter for the two frameworks in question, related to issues one through five.

ROOT-CAUSE RESULT FOR LACK OF WELL-ESTABLISHED REQUIREMENTS	Addressed	Not Addressed	Partially Addressed
1 Lack of Established Project Objectives			X
2 Missing Stakeholders			X
3 Poor Communication with Stakeholders			X
4 Users Don't Know What They Want		X	
5 Errors of Omission		X	

ROOT-CAUSE RESULT FOR LACK OF WELL-ESTABLISHED REQUIREMENTS	Addressed	Not Addressed	Partially Addressed
5.1 Lack Of Solid Process on How to Gather Requirements		X	

Table 32. Summary of PMBOK® Framework versus Issues 1 to 5 of Lack of Well-Established Requirements

ROOT-CAUSE RESULT FOR LACK OF WELL-ESTABLISHED REQUIREMENTS	Addressed	Not Addressed	Partially Addressed
1 Lack of Established Project Objectives		X	
2 Missing Stakeholders	X		
3 Poor Communication with Stakeholders	X		
4 Users Don't Know What They Want			X
5 Errors of Omission		X	
5.1 Lack of Solid Process on How to Gather Requirements		X	

Table 33. Summary of SE Framework versus Issues 1 to 5 of Lack of Well-Established Requirements

As can be inferred by the analysis above, the SE framework is the one that provides an answer for two of the issues identified. However, neither framework has a complete answer to address all of them.

For thoroughness, we should also cross-reference this engagement model with the poor planning root causes. The summary of such root causes is presented in table 34 for convenience.

ROOT-CAUSE RESULT FOR POOR PROJECT PLANNING
1 Lack of Established Project Objectives
1.1 Poor Communication with Stakeholders
1.2 Missing Stakeholders
1.3 Lack of High-Level Scope of Work
1.4 Lack of Acceptance Criteria
2 Poor Risk Identification
2.1 Unbalanced Business and Technical Risk Identification
2.2 Not All Appropriate Stakeholders Involved in Risk Identification
2.3 Project Manager Tries to Do It by Herself
2.4 Risk Register Is Not Made a Living Document
2.5 No Formal Risk Identification Process Done as Part of the Pre-contract Phase by Service Organizations
2.6 Lack of Understanding of the Importance of Risk Identification
3 Lack of Validated Assumptions
4 Lack of Stakeholder Buy-In to the Plan

Table 34. Identified Root-Cause Items for Poor Planning

Besides the root causes that are common to both lack of well-established requirements and poor planning that were mentioned during the analysis performed above, lack of well-established project objectives, missing stakeholders, and poor communication with stakeholders, there are other root causes of poor planning that can be cross-referenced against the two main underlying issues with this type of integrator engagement.

Obviously, the missing stakeholders issue identified in the previous analysis also drives issue 2.2 in the table above, not all appropriate stakeholders involved in risk identification. Lastly, the fact that the integrator doesn't necessarily have the full picture on the business value of the test system for the client generates an immediate imbalance on the technical and business risk identification, our issue 2.1 in the table above.

Tables 35 and 36 present the gap analysis of both frameworks against issues 2.1 and 2.2 identified above.

ROOT-CAUSE RESULT FOR POOR PROJECT PLANNING	Addressed	Not Addressed	Partially Addressed
2 Poor Risk Identification			X
2.1 Unbalanced Business and Technical Risk Identification		X	
2.2 Not All Appropriate Stakeholders Involved in Risk Identification			X

Table 35. Summary of PM Framework versus Issues 2.1 and 2.2

ROOT-CAUSE RESULT FOR POOR PROJECT PLANNING	Addressed	Not Addressed	Partially Addressed
2 Poor Risk Identification			X
2.1 Unbalanced Business and Technical Risk Identification		X	
2.2 Not All Appropriate Stakeholders Involved in Risk Identification			X

Table 36. Summary of SE Framework versus Issues 2.1 and 2.2

Though both frameworks address the "not all stakeholders involved in risk identification" issue, they still leave gaps in the imbalance of business and technical risks.

As can be seen by the cross-referencing analysis presented in this section, the proper application of each or both of the two frameworks in the execution of the T&M system by the integrator where the client fully delegated the initial requirements definition to the integrator is certainly not enough to mitigate the problems that it brings.

Ideally, the engagement with the integrator needs to be changed to one where it is not fully in charge of collecting the project requirements. The client needs to take responsibility for that activity, as it is definitely in the best position to make sure the full business value for the test system is being taken into account. As we saw previously in this text, though, some clients are not well equipped to do so either by not having the needed resources on staff or by suffering from overallocation of said resources to other projects.

What this text suggests is that since this activity is a driver of project success, it deserves full focus. The idea here is to either hire extra staff to execute it, or to hire a consultant with the appropriate experience in the system integration business, technical skills, and business savvy to be the voice of the client and interface with the integrator on this activity.

Client Does Back of Napkin Requirements Definition

The second scenario that was described in the last chapter as a challenge in having system integrators involved in building a T&M system for clients happens when clients do an abbreviated requirements definition exercise prior to integrator engagement.

As detailed in the last chapter, there are some facts usually present in this type of engagement with integrators. Those facts are repeated here for convenience:

- The client may not be as well versed in performing a thorough requirements analysis for T&M system as an expert from the field would be
- Common for internal resources not to have in-depth knowledge of the type of information an integrator usually needs from clients in order to keep headed in the right direction
- Internal resources rarely can dedicate themselves full time to the requirements-definition exercise
- It is usually extremely difficult to find resources who can execute a well-balanced business and technical requirements-gathering activity

If the client is not as experienced in performing requirements analysis for T&M projects, even though the integrator will be starting off from a more advanced stage than in the scenario where the entire requirements definition is performed by that organization, the incomplete requirement set generated by the client will need to be advanced into a more comprehensive requirements set by the integrator. This means the integrator will probably have to take over the requirements-gathering exercise at that stage.

This is also valid for the case when the client doesn't have an in-depth knowledge of the type of information needed by integrators to increase odds of success. The integrator will not only have to take over the requirements-elicitation exercise, but will probably even need to go back to collect the project objectives from the client.

It is easy to see how these first two items from the list above would drive the engagement toward the first one described, where the integrator would end up taking over the requirements-definition efforts.

Going down the list of issues, if the internal resources can't dedicate themselves full time to the requirements-elicitation efforts, either the integrator will have to take over for them or the project will start with insufficient time spent on requirements gathering. Regardless of how the scenario ends up, this engagement will present all the symptoms of the prior engagement where the integrator is made responsible for the requirements-elicitation activity.

Lastly, it is usually difficult to find resources that can be well versed in T&M business and technical aspects as well as having the business savvy to make sure the business requirements are gathered. This also resembles the scenario where the integrator is fully in charge of the requirements elicitation. It will create an imbalance of business and technical requirements, with the consequences we have seen previously in this text.

In summary, this type of engagement can very easily bring the same consequences as the previous one presented. Therefore, the same analysis performed for issue one, the scenario where the integrator is fully in charge of collecting the project requirements, is applicable to this issue as well.

Over the Fence Mentality

The next scenario that was described in the last chapter as a challenge in having system integrators executing T&M systems for clients happens when clients have an "over the fence mentality." As presented, the over the fence mentality happens when the client believes that since an expert company is being hired to execute the T&M system, the system integrator will take care of everything with absolutely no or minimum interaction with client internal resources until it is time to deploy the system.

This engagement model actually brings several subtleties when compared against the two main root causes of T&M project failure.

Let's start our cross-referencing exercise with the issue "integrator focuses only on system risks (maybe)" presented above. Much like in the two previous engagement types presented in this chapter, the focus solely on the test system's technical risks will cause an unbalance between the technical and business risks, root cause 2.1 of the poor planning summary table.

Another item that shows in this engagement type that is also common to the previous two engagement types is the lack of understanding on the integrator's part of the full business needs by the client for a given test system. We can borrow the analysis performed in the previous engagement and cross-reference this issue with root cause number one, lack of established project objectives, for both PMBOK® and SE frameworks.

The fact that the client may not have resources to closely manage the progress made by the integrator is an issue with far-reaching consequences. First of all, it is easy to understand that a project that runs open loops with its stakeholders can easily suffer from errors of omission. It was mentioned previously that the requirements-gathering process is an iterative process. As such, it requires multiple iterations with the project stakeholders and constant refinement. In fact, it is unrealistic to expect a pure waterfall execution for a T&M system. Usually, the farther into the project life cycle the project is, the more the stakeholders and integrator understand the true nature of requirements and how these requirements may need to be adjusted in order for the system to capture full business value.

If there is no closed loop with the client's stakeholders, the project requirements may never get adjusted until the "oh-oh" moment at deployment, when the client's stakeholders may realize that what is being delivered is not really what the client needs—errors of omission during the project execution due to lack of involvement from the client. As seen before, "users don't know what they want" is a corollary to the errors of omission root cause, and it is present in this case also, as described above.

Furthermore, this lack of iterations between client and integrator, as seen on the analysis for the previous engagement type, also drives missing stakeholders and poor communication with stakeholders, much like in the previous engagement type.

One last note that is important to capture about this engagement type is in regard to the "users may not have technical expertise to follow the implementation" issue. This lack of involvement on the technical decisions made by the integrator on how to solve the problem drives a lack of buy-in to the project plan.

The project plan is basically the blueprint showing how the proposed problem, the test system, will be solved or implemented. If the client doesn't have the needed technical expertise to follow what is being proposed by the integrator, true project plan buy-in will never be achieved, and, consequently, there may be an us-versus-them relationship between the integrator and client.

The following tables summarize the cross-referencing analysis of the two frameworks considered in this text against the issues surfaced by this engagement type.

ROOT-CAUSE RESULT FOR LACK OF WELL-ESTABLISHED REQUIREMENTS	Addressed	Not Addressed	Partially Addressed
1 Lack of Established Project Objectives			X
2 Missing Stakeholders			X
3 Poor Communication with Stakeholders			X
4 Users Don't Know What They Want		X	
5 Errors of Omission		X	

Table 37. Summary of PMBOK® Framework versus Issues 1 to 5 of Lack of Well-Established Requirements

ROOT-CAUSE RESULT FOR LACK OF WELL-ESTABLISHED REQUIREMENTS	Addressed	Not Addressed	Partially Addressed
1 Lack of Established Project Objectives		X	
2 Missing Stakeholders	X		
3 Poor Communication with Stakeholders	X		
4 Users Don't Know What They Want			X
5 Errors of Omission		X	

Table 38. Summary of SE Framework versus Issues 1 to 5 of Lack of Well-Established Requirements

ROOT-CAUSE RESULT FOR POOR PROJECT PLANNING	Addressed	Not Addressed	Partially Addressed
2 Poor Risk Identification			X
2.1 Unbalanced Business and Technical Risk Identification		X	
2.2 Not All Appropriate Stakeholders Involved in Risk Identification			X
4 Lack of Stakeholder Buy-In to the Plan	X		

Table 39. Summary of PMBOK® Framework versus Issues 2.1, 2.2, and 4

ROOT-CAUSE RESULT FOR POOR PROJECT PLANNING	Addressed	Not Addressed	Partially Addressed
2 Poor Risk Identification			X
2.1 Unbalanced Business and Technical Risk Identification		X	
2.2 Not All Appropriate Stakeholders Involved in Risk Identification			X
4 Lack of Stakeholder Buy-In to the Plan		X	

Table 40. Summary of SE Framework versus Issues 2.1, 2.2, and 4

As the cross-referencing analysis above shows, the only issue that is properly addressed by one of the frameworks is the lack of stakeholder buy-in to the plan. However, in this scenario where the client just doesn't have the needed technical resources to go through the buy-in process, applying the correct framework will do little good to the overall project success. This is a deeper issue that needs to be addressed, in fact, by the client, either by having an unbiased consultant that is not involved with the integrator executing the project to function as that technical liaison or by actually hiring that technical expert to be part of its staff and execute that role.

The better option between the two varies on a case-by-case basis. If the client foresees several of projects down the line where it will utilize integrators, it may make sense to invest on a full-time hire to fulfill that role. If the client's business is such where these types of projects come few and far between, the consultant route may make the most sense.

T&M System Implemented in Parallel to DUT Design

The next engagement type, described in the last chapter as a challenge in having system integrators executing T&M system for clients, happens when the device under test is being design in parallel to the actual system to test it. Figure 21 summarizes the consequences of this challenge.

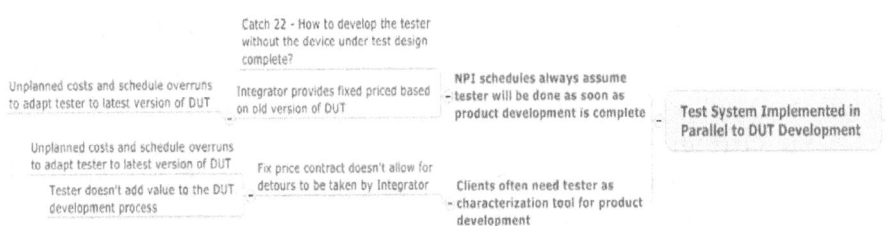

Figure 21. Issues with Test Being Developed in
Parallel to DUT Development

This engagement type is probably the one where there is the biggest disconnect between the engagement of integrators and the business value to be captured by the T&M system.

As mentioned previously, historically, integrators have been applying a typical waterfall approach to executing T&M projects. Also, as seen before in this text, waterfall development emphasizes more time spent up front in the project life cycle, highlighting the importance of requirements and design phases to complete, prior to the implementation phase to start. As integrators are asked by clients to provide fixed-price proposals for the execution of T&M systems, the easiest way for them to have a level of confidence that the risks are somewhat known and under control is to make sure the requirements are known to a somewhat high level of certainty, before they start project execution.

The problems summarized by figure 21 are the by-product of this behavior. In fact, an analysis of the two main frameworks utilized by integrators to execute T&M systems is not even applicable in this case, as they both use waterfall development as a foundation.

What this book suggests is the utilization of a modified agile framework with staged delivery for situations like this. Section 2 of this text includes a section dedicated to detailing how such methodology can be applied to increase the odds of success of this engagement type. For the time being, let's conclude the analysis for this engagement type by stating that this is one of the most challenging scenarios in the engagement between clients and integrators due to the historical nature on what is expected from both sides.

The integrator expects to be able to develop and/or receive a clean-cut, immutable set of project requirements that can be used as a basis for the development of a fixed-price proposal with schedule, price, and scope of work. The expectation is that this initial set won't be changed, or if changed, a well-defined set of changes can be identified and incorporated on change orders to change the original terms of the contract.

The client expects to receive a fixed-price proposal with spelled-out and immutable project schedule, price, and scope of work at the same time it expects the final system to be completed, practically at the same time the unit under test will be ready for market product launch.

Both sides are wrong in their assumptions. The NPI (new product introduction) process is, by nature, very fluid until its completion with the market launch. Changes throughout the entire NPI life cycle may not be, and most likely won't be, clean-cut and easily packageable into change orders by the integrator. There has to be some flexibility in the way the contract was structured so the integrator can still make its money on the arrangement, even if change orders are difficult in generating. On the flip side of this coin, the client needs to understand that in order for the best business value for the test system to be captured by the organization, the initial project budget and schedule should account for the fact that changes will happen along the way and that change orders will be generated to capture them into the original contract.

Project Budget Defined Before Feature Set Complete

The last engagement type described in the last chapter as a challenge in having system integrators executing T&M system for clients happens when the project budget for a given T&M project is defined by the client's organization ahead of the actual feature set for the system is complete. Figure 22 summarizes the consequences of this challenge.

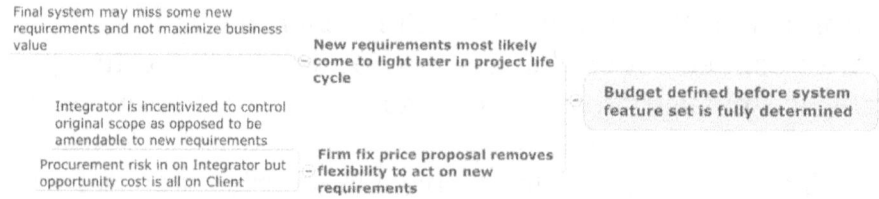

Final system may miss some new
requirements and not maximize business
value

New requirements most likely
come to light later in project life
cycle

Budget defined before system
feature set is fully determined

Integrator is incentivized to control
original scope as opposed to be
amendable to new requirements

Firm fix price proposal removes
flexibility to act on new
requirements

Procurement risk in on Integrator but
opportunity cost is all on Client

Figure 22. Issues with Budget Defined Before Feature Set Complete

This last engagement type is a very close corollary to the previous one presented. The reason why this one is presented as a separate type is due to the fact that not only NPIs suffer from this problem.

As presented in the previous engagement type, typical NPIs call out for a test schedule and budget. This is passed onto the T&M project as its initial constraints so the test development fits the overall NPI parameters. At that point, usually the device under test is not ready by the client's product development group. What that means is that the final T&M project requirements set cannot really be defined ahead of a proposal to be generated for the system.

Several client organizations follow the same methodology for the outsourcing of their T&M projects, even if they fall outside an NPI process. These organizations usually don't invest the appropriate amount of time in defining the T&M system feature set prior to engaging integrators on the procurement process.

This engagement type is the combination of all problems that were presented on the previous engagement types, as it brings characteristics of all of them onto itself.

On top of all that was presented in this chapter as potential issues that can come about when clients engage integrators to execute test and measurements projects, professional project management is unfortunately not commonplace among system integrators. The project management discipline, though making excellent progress in areas such as IT and construction, is still crawling its first steps in the test and measurements industry. It is not common to find professional project managers certified by PMI or even professional systems engineers certified by INCOSE as part of system integrator's staff.

With this information in hand, clients can make more informed decisions on how to select integrators to build their T&M systems. What it is being proposed here is that the old selection criteria of integrators by clients based on proposed price, implementation schedule, and technical ability are no longer enough to increase the odds of project success. With today's increased T&M project complexity, project management and peripheral business vision on how the test system will fit into the client's business are, if not more, at least as important as the three criteria points above. These two aspects of the relationship between clients and integrators should definitely be taken into consideration as part of the procurement process and contract awarding.

Long gone is the time when a client would throw a set of test system specifications over the wall to an integrator and receive a perfect test system back from the other side. Today's challenges are forcing a deeper partnership between client and integrator. The client needs to make sure it is taking ownership of the project success as much as the integrator that is executing it. Again, it is in the client's best interest to make sure the T&M project best aligns with the business objectives.

What it is being proposed here is the creation of a role to fill the gaps presented in this chapter. The person in this role should understand the system integration business to help the client help the integrator in that partnership. This person should also be skilled in the technical aspects of the test and measurements industry, be a strong professional project manager, and have enough overall business savvy to make sure the client's business objectives are always kept in perspective while the T&M project is being executed.

Furthermore, as seen in the presented analysis so far, both the PMBOK® and SE frameworks alone are not the answer to increase T&M project success. Both frameworks have strengths and weaknesses when applied to T&M projects. One initial idea is to combine the strengths of both into a process that is better tailored for T&M projects. The analysis performed so far has set the stage for that process to be presented in section 2.

It is also important to highlight that any process will only be as good as the people implementing it. With that in mind, it is important for the reader to always keep the order of importance of people, process, and tools in this order. People are always the most important aspect of any project team. Good people become even better if empowered by a strong process. Processes become more flexible if composed of a set of good tools.

Section 2 of this book will focus on:
- A proposal for the structure of the project team to maximize odds of success as well as roles that should be filled when an integrator is involved in building a system for a client
- A proposal for a hybrid process tailored for T&M projects
- The presentation of tools that should be leveraged by the process

CHAPTER 5:
Problem Statement Summary

Section 1 of this book was devoted to presenting the major issues that contribute to why T&M projects fail. The section started off by presenting an analysis that was based on years of project postmortem data collection on T&M projects, in search for the root causes of the low odds for success of such projects. The analysis culminated into the identification of the underlying issues for each one of the two major root causes for T&M project failure: poor planning and lack of well-formed requirements. As system integrator companies are often involved in the execution of T&M projects, the text progressed to the root-cause analysis for why T&M projects fail in general with the presentation of the five main pitfalls in working with companies such as these when building T&M systems.

With the stage set and a clear picture painted on the main issues that needed to be addressed, in both executing T&M projects directly or involving system integrator companies to do so, the book focused on executing an cross-referencing analysis of strengths and weaknesses of the two main project management frameworks for the execution of T&M projects against the issues identified and presented in the previous two chapters. The attempt behind such analysis was to identify whether the reasons why T&M projects fail were due to lack of training or correct application of these frameworks, or if the frameworks themselves were insufficient to increase the odds of success for T&M projects. As it was seen in that chapter, the two frameworks are not sufficient if applied in a standalone mode and without adaptation of their standard form to yield better results.

Lastly, the section executed the cross-referencing of the two frameworks against the identified issues in working with system integrator companies to build T&M projects. The goal for this cross-referencing was to check if these pitfalls could be minimized by the appropriate application of the two main project management frameworks, and if not, to identify the gaps on these frameworks that needed to be filled in order for this relationship to provide overall better results. As was seen, not only were the frameworks not enough in themselves to improve the relationship between integrators and clients, but also the need existed for both a cultural change in the mind-set of clients and integrators as well as a missing role in the relationship.

Now that the problem statement has been properly formulated, section 2 will focus on the proposed solution for the identified issues. The text will focus on changes to the people, process, and tools combo and propose a new framework for the execution of T&M projects. The new framework proposed will focus on each element of the people, process, and tool set and provide best practices that were successfully applied in the execution of several T&M projects over the years. The framework will also focus on the client-integrator relationship and try to address the five main issues identified in section 1 of this text.

SECTION II:
The TMPM Framework

CHAPTER 6:
The TMPM Framework

The first chapter of this section focuses on the people element of the people, process, and tools combo. In order for the case to be made for the organization structure that will be proposed, it is useful first to create a foundation for the analysis. The foundation will be twofold: the first half will present the typical anatomy of a T&M project team as well as an example of the types of stakeholders that will interface with this project team in a typical T&M project, and the second half will go deeper into the root causes that were identified and presented in section 1.

As mentioned previously, today's technical requirements for T&M systems have reached an unprecedented level of complexity. T&M systems today require a much higher number of engineering disciplines and specialties involved in their development when compared to their counterparts of the past.

Let's take a simple illustrative example of a test system to test cell phones and derive a likely anatomy of a project team to create such system.

For starters, a strong test engineer is obviously needed. A test engineer is a professional skilled in the creation of a process that will test a given unit in several different scenarios, such as manufacturing, product development characterization, quality assurance, and RMA (return merchandise authorization). This professional is the one who is usually best suited to think through the best way to architect the test specifications for a product in a way that can maximize the return of capital investment by the organization, depending on the target utilization of such a test system. This person is also usually responsible for determining the best way a test can be performed in order to achieve maximum test coverage.

In the past, test engineers would have all, or at least a good chunk, of the needed skills in software and hardware to take on the complete, or very close to complete, design and implementation of the final test system solution along with the responsibilities listed above.

As products grew in complexity, their corresponding test systems followed that pattern and became true engineering marvels. The level of technical intricacy of a cell phone test system is orders of magnitude higher than a test system for a washing machine or microwave, for instance.

Not only that, but in the information age, the expectation is that the test data is to be available and accessible by other corporate systems, such as MES (manufacturing enterprise system), ERP (enterprise resource planning), MRP (manufacturing resource planning), SPC (statistical process control), etc. Also, it is highly desirable that reports be made available online through a web browser, not only on the organization's intranet but external to it.

Databases have been migrating onto the so-called NoSQL or schemaless databases where the level of flexibility for data meshing and integration with other databases is much higher as is support for the so-called big data revolution.

Consequently, test data storage has became much more than simply dumping data into a relational database and supporting a few queries for data retrieval and reporting. It has migrated into a very complex information technology problem. The same idea can be extrapolated to all areas of test software development. Nowadays, the needed depth of software skills for a T&M project team has increased in all directions, to a point that is virtually impossible for the test engineer to do it all.

With this in mind, the project team definitely needs software professionals. These people make careers around becoming extremely proficient and vertically knowledgeable in the discipline of software engineering.

Going down the other direction in software skills, modern units under test usually have some sort of test firmware needed in order to verify their application firmware and embedded hardware. Even though firmware software can be technically considered software engineering, the required skills by an engineer who designs and implements firmware software is totally different than the skills of a software engineer who gravitates toward data storage, web programming, and enterprise-level software. Therefore, a firmware engineer needs to be included as another member of our hypothetical cell phone test system project.

Since we mentioned firmware engineer as a needed member on our team, it can be implied that this firmware will run in some sort of embedded hardware platform. Therefore, an embedded engineer is another person we will need to include on our project team.

Since we are migrating now to the hardware aspect of our cell phone test system, it is important to think about how the cell phone signals will get to the test instrumentation. Even though off-the-shelf instrumentation has evolved tremendously in the past few years, more often than not, signals of interest coming from the unit under test need to be conditioned prior to being measured by the off-the-shelf instrumentation.

This has always been the case; however, the complexity of the unit under test hardware and its signals of interest have followed the same trends as mentioned previously in this chapter. What used to be just a matter of conditioning signal levels with some operational amplifiers, for our cell phone case, it becomes signal conditioning of extremely high frequency signals.

This type of signal conditioning presents several challenges, ranging from extra care that needs to be taken when laying out the printed circuit board that will house the electronic components of the signal conditioning board, all the way through the higher impact of electromagnetic interference when the circuit operates in the RF range. This suggests that this task, which has been historically handled by strong test engineers, now requires a professional with vertical knowledge in the electrical engineering arena.

Still on the hardware side of things, miniaturization of devices has become almost an obsession in modern society. Using our cell phone case as an example, the cell phone of twenty years ago was probably ten times as heavy and three times as large as modern phones. This suggests that mechanical fixturing to provide physical connectivity and housing of the unit under test has also become much more challenging. A strong mechanical engineer is therefore a must-have on our project team.

As cellular telephony has made its strides, there is now much tighter regulation of frequency bands. In the interest of facilitating such control, standards were developed. The verification that devices are following those standards is another requirement that must not be forgotten when developing the test system. As such, a validation engineer is also a very important person to be included as part of our project team. This person will make sure the standards are being met and the needed paperwork is being generated in order for the organization to obtain product approval by the corresponding regulating agencies.

Since the cell phone test system is obviously an RF project, we must not forget that a strong RF engineer will probably be the only one who can figure out those hairy problems that happen in the gigahertz range. Therefore, let's include an RF engineer on our project team.

Lastly, let's not forget our project manager, the person who will supposedly coordinate the work of all these team members into a cohesive technical implementation.

Based on what we have so far, the figure below represents the basic structure for our project team.

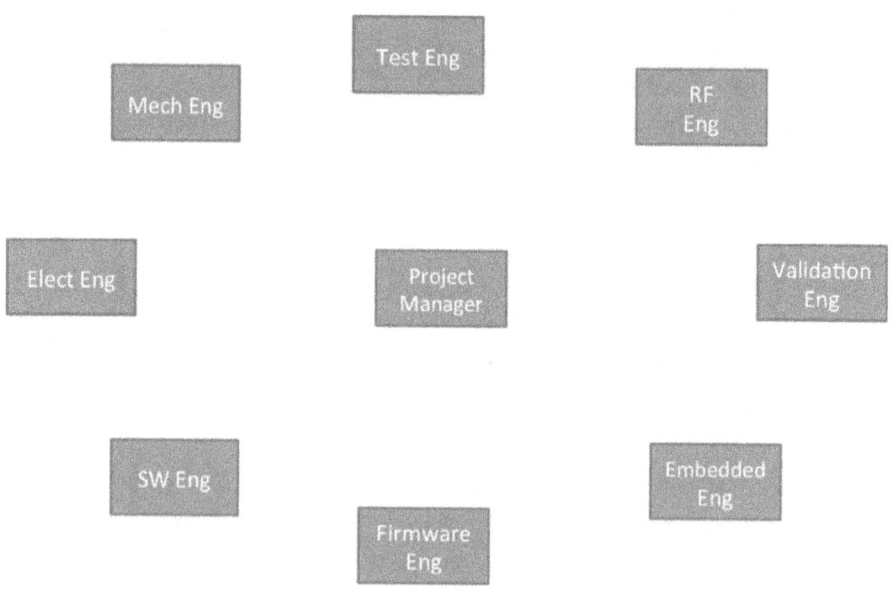

Figure 23. Typical T&M Project Team Structure

Even though we are making our case by using the example of a cell phone tester, the point can be made for other types of units under test. The take-away point for this initial discussion is that the project manager is expected to have enough technical depth to intelligently interface with all the technical players and the different disciplines of expertise that are involved in the project team's dynamics.

Let's keep this analysis in the back of our minds for now and switch gears to another level of interaction that involves the project manager: the project stakeholders. Assume that the cell phone being tested with a system built by our project team is part of an NPI (new product introduction) for a product that is expected to revolutionize the cell phone market. This system is expected to be a production test system that will perform manufacturing quality assurance for the units prior to shipment from the factories to the stores.

The most obvious stakeholder that will need to have her voice heard is the test operator. The test operator will be the primary user of the test system once deployed to production. The operators can have multiple different backgrounds, which may or may not college education. This person, however, is usually extremely experienced on problems that prior test systems for similar units under test presented and is definitely one who should be heard by the project manager, on both requirements gathering and risk identification.

Since this cell phone is part of an NPI, it means it is being developed in parallel to the test system project. Obviously, the project manager ought to include the product development engineers in the project stakeholder registry and make sure they have plenty of opportunities to voice their opinions on requirements gathering, risk identification, and risk analysis. The backgrounds from this group will probably be close to the ones of the project team members, extremely technical in various disciplines.

Since the cell phone being showcased is a new release, the organization will definitely have a product manager. This person is responsible in leading the charge on the features to be included as part of the initial release of the product. Every NPI process is very fluid when it comes to feature set. There are several variables that are considered by an organization when determining the feature set that will be included in an NPI launch, such as schedule for launch, size of prerevenue investment, market research, data from focus groups, analysis of competitors, target sale price, etc.

Since the marketplace is always changing, the direct reflection on an NPI process is that the feature set is maintained in a state of flux for a long period, which usually overlaps with the start of the test system project. As such, the test system project manager should definitely include the product manager in the stakeholder registry so changes in the final feature set are captured and reacted to by the test system project team. Product managers have different backgrounds depending on the industry they serve. In our illustration case, this person will probably have a technical background plus some sort of business education.

Since the test system will be a production system to ensure the quality of the shipped units, the organization's quality manager should also be included in the list of stakeholders for this project. This person will have important input on the requirements for the test system and the KPIs (key performance indicators) that will be important to control production. Quality managers also have multiple backgrounds. In our example, this person will probably either have a technical background or some sort of regulatory experience.

The test system will be a capital expense for the organization. This means there most likely should be a procurement agent involved in the list of stakeholders. Especially in large organizations where purchases need to flow through the procurement department, the project manager should make sure to include this person, at a minimum, in the risk identification and risk analysis meetings to make sure there will be no surprises in procurement that can potentially derail project cost and schedule. Procurement agents usually have some sort of finance background.

As the system will be deployed from the production floor, once deployed, it will become the responsibility of a production director. This person understands and cares about yields, production capacity, test times, and other production-related KPIs. As such, it is paramount that this person is involved in requirements-gathering activities so the project manager understands what types of production support numbers the overall project will need to achieve. Production directors usually have some sort of industrial or manufacturing engineering background.

Depending on the level of visibility of the NPI on large organizations, and probably in all cases in smaller companies, there will be some sort of C-level (CEO, CTO, COO, etc.) person who should be involved as a stakeholder. This person will provide an escalation path for the difficult decisions and a voice of command that will rally the troops to march in the same direction throughout the entire project execution. The C-types usually understand the language of money more than any other, so a finance background is certainly helpful when interfacing with them.

Figure 24 shows the multiple stakeholders that the project manager must interface with and their corresponding backgrounds.

As one can see, the list of people the T&M project manager needs to interface with is usually quite extensive. Moreover, the backgrounds of these people are quite diverse, ranging from deep vertical technical expertise, through accounting, finance, manufacturing engineering, and all the way to the street smarts of the production operators.

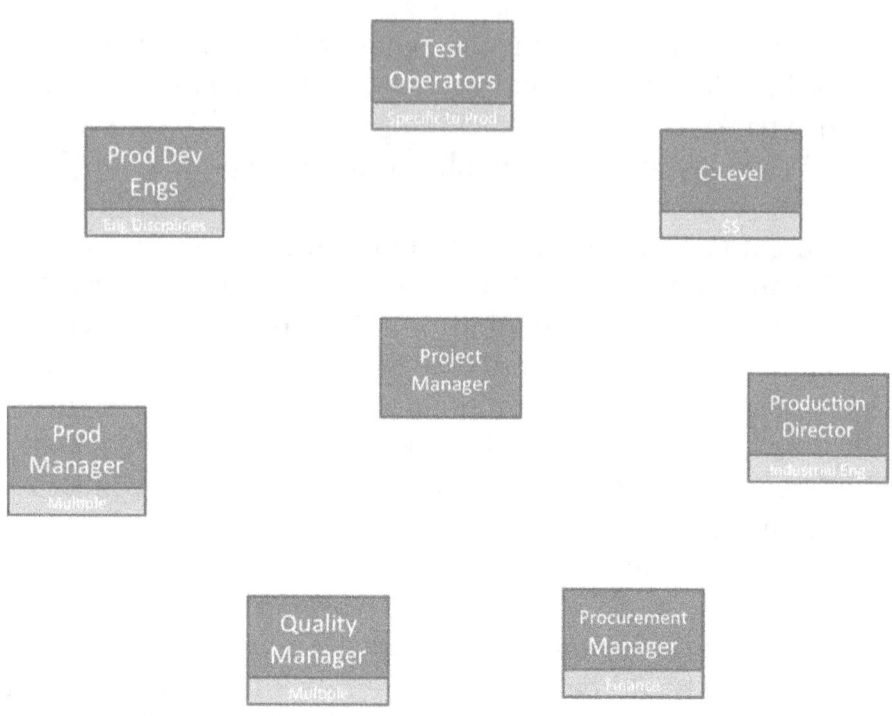

Figure 24. Stakeholders and Corresponding Backgrounds

Let's now switch gears to the second half of our target analysis, the root-cause analysis of section 1. We shall start this portion of our analysis by looking at the underlying issues that were identified as part of poor planning, repeated here for convenience.

Project objectives can be seen as the highest level of abstraction of the project requirements. In fact, the project objectives will be the source from which the requirements gatherer will start its activities of collecting the project requirements from. In an NPI process, it is easy for one to see how the project objectives should capture the high-level "care-abouts" for the project stakeholders.

Figure 25. Poor Planning Root Cause: Underlying Issues

For instance, it is important that project objectives capture the production KPIs that will be used by the production director to control production of the new device. The details around these KPIs will unfold as part of the requirements-elicitation exercise that will follow the collection of project objectives. However, if the project objectives fail to mention the production KPIs, the person collecting requirements later may very easily flat out neglect to focus on that area of the project requirements.

Another easy-to-grasp example would be the list of features that are currently included as part of the NPI launch, provided by the product manager. Each one of these features will unfold into several lower-level project requirements; therefore, each one will be use as a blueprint for the requirements collection around testability of product features. If they are neglected, it is almost a guarantee that some testability requirement will be missed, especially in larger projects.

The concept of garbage in, garbage out is very much a reality when it comes to project objectives and project requirements. Also, it is of extreme importance that these project objectives come from the project stakeholders and are not filled in by the project team members or project manager.

The project manager should avoid the trap of filling the gaps with her own assumptions, beliefs, and background. Each project stakeholder is the most skilled and knowledgeable person to provide project objectives in their own areas of domain. For instance, the project manager should not assume the KPIs that will be used to control production; they should be collected from the production director. This type of filling the blank activity is a true recipe for disaster, regardless of how experienced and knowledgeable the project manager is. A missed production KPI, for example, can snowball into several missing requirements, which will lead to a bad system design, which will lead to errors of omission in the implementation, which will lead to rework/redesign of the solution at the very end of the project, the most costly of all types of rework.

Not only that, but missed project objectives that lead to missed requirements will definitely lead to poor project planning as the planning is done based on the project requirements as one of its inputs.

Since the project objectives need to come from the project stakeholders, it is extremely important that whoever is responsible to collect them can connect and communicate well with all stakeholders. If the communication is faulty, due to the fact that the backgrounds of the people involved in the communication are so different that there is a huge gap in knowledge preventing the interchange to be effective, something will fall through the cracks. If the project stakeholders and the person collecting the project objectives don't speak the same language and don't fully comprehend each other, objectives will either be missed altogether or will be collected in a manner that don't correctly represent the intent behind the objectives.

As seen in the first half of our analysis, a T&M project involves several stakeholders with multiple backgrounds. Therefore, it is of paramount importance that whoever is collecting the project objectives properly connects with all stakeholders. And there we have the first huge gap in the way T&M projects have been executed. Usually, there is a project manager who is the person who executes this task of collecting project objectives. In order for this project manager to connect with all these stakeholders and their multiple backgrounds, this single person would need to have a level of understanding in an incredible number of different disciplines in order to provide full coverage. It is virtually impossible for any single individual to be fluent in all different types of technical skills that are required in complex T&M projects at the same time, who is also skilled in manufacturing, finance, accounting, business, and also has a way of connecting with people that allow her to properly communicate with both C-level stakeholders and operators, the two extremes in an organization.

The next underlying issue is poor risk identification. It was previously mentioned that the stakeholders should definitely be involved in risk identification and risk analysis for any project. They are in the best position to point out risks in their own areas of expertise, much better than someone who doesn't do their jobs for a living.

It was also mentioned in a previous chapter that projects having a higher risk ratio end up failing. Risk ratio was defined as the ratio between the number of unforeseen risks divided by the number of identified risks.

Now, if the person conducting the risk identification exercise can't properly communicate with the multiple stakeholders and their different backgrounds, odds are that several risks will not be identified or become known risks. This will lead the risk ratio to be high, the kiss of death for T&M projects.

It was mentioned previously that the requirements-gathering exercise is an iterative task, where the more the requirements evolve and the more stakeholders understand them, the more often these stakeholders suggest that other stakeholders be involved in the activity as these other people supposedly can better identify some given requirements that became visible. This is also true for risk identification. The more the risk registry evolves and the stakeholders are involved in the process, the more they understand the exercise and can point the risk gatherer toward stakeholders that were originally missing from the stakeholder registry.

Obviously, missing stakeholders could have identified other risks that were not identified by the original stakeholder list. This will again lead to an increase in the risk ratio.

The next underlying issue down the list is lack of validated assumptions. This is another very important topic that should not be ignored. As it was mentioned before in this text, test systems very often need to start before the unit under test has been completely designed, typical of NPI processes. In other cases, the requirements-gathering process wasn't done as thoroughly as needed prior to the project implementation starts. For both cases, the only way for the project team to actually start on the T&M project is by stating a list of assumptions the team will use to implement the system.

It is important that these assumptions are captured so the project team and project stakeholders understand the framework the project is based on, so progress can be made. Also, these assumptions will be used as input for the project planning.

Equally as important as capturing the assumption is their validation by the project stakeholders. Again, the project stakeholders are the most capable people to validate the assumptions that are within their areas of expertise and proper to their business, making sure the project team is not starting the project off with bad assumptions. Bad assumptions can very easily drive the project team into a hole that would be very costly and time consuming to climb out of.

If the process of communicating with the project stakeholders is defective, they most likely won't fully understand what the assumptions mean, or the person collecting assumptions from them will not correctly translate the assumptions to the project documentation, or they will not be fully engaged in the process. The same iterative process where project stakeholders that fully grasp the project assumptions refer the project manager to missing stakeholders that have something to contribute toward the project assumptions is also valid. This contributes to the identification of missing stakeholders that were not originally included in the stakeholder registry. One more time, the communication with the project stakeholder is at the root of the problem.

The last underlying issue for the poor planning root cause is lack of stakeholder buy-in to the project plan.

The project plan is basically the blueprint showing how the project will be carried out. A T&M project is basically an engineering solution to a problem. Therefore, the project plan is basically the definition on how the proposed problem will be solved.

There is an area of psychology called change management, which is relevant to this discussion. Change management, as defined by Wikipedia, is an approach to transitioning individuals, teams, and organizations to a desired future state. One of the most important tenets of change management is the fact that the person driving the implementation of a change should not try to sell the change to people who will be affected by it, as a way of accelerating agreement and implementation.

Human beings respond better to change if they are part of the change themselves, meaning that they participated in the process of defining what changes were needed in the first place. In other words, people prefer to participate in the architecture of the solution rather than be presented with the solution that was predetermined by other people. The individuals who participate in the process of defining the change are much more amiable to the change process, and often become advocates for the change themselves, champions of the process per se.

Obviously, changes are only contemplated when there is a problem that needs solving. No organization, or even a person, would go through change if not motivated by the desire or need to solve a perceived problem. In other words, change management is a solution to a problem, much like the project plan for a T&M project is a solution to a perceived engineering problem.

Using the ideas presented by change management, if the project stakeholders are sold on a project plan, their buying-in will not have the same level of commitment as if they were involved in the creation of the project plan. Obviously, it is not realistic to expect that all project stakeholders, from the test operators all the way up to the C-level stakeholders, will have a say at every single nuance of the project plan. However, it is best practice to involve them in the areas of the project plan that touch their direct areas of expertise.

T&M projects are extremely challenging for multiple reasons. One certainty in life, like taxes and death, is that problems will happen on complex T&M projects. No complex T&M project is smooth sailing from beginning to end. If the project stakeholders truly buy into the project plan, to the extent they became advocates of the proposed solution, they will be much more agreeable to helping in times of need.

They will most likely have a totally different attitude toward the project manager of a T&M project that made them feel they were part of the solution than toward one that arrogantly shoved a project plan down their throats. In the former scenario, an environment of cooperation and understanding will probably be created around the project team. In the latter scenario, there may be an I-told-you-so attitude from some stakeholders that adds absolutely no value toward project success.

All of this is only possible if the communication with the project stakeholders is stellar. If they don't understand what this thingumajig is supposed to do, they will most likely not be fully bought into the proposed project plan, and the latter case described above will unfold.

For thoroughness, the analysis will turn to the underlying issues for the lack of well-established project requirements root cause. Figure 26 shows the top five underlying issues.

The lack of project objectives underlying issue to bad requirements is also common to poor planning, as presented in the previous analysis. It was highlighted how the project objectives are a higher level of abstraction to the project requirements, the latter being a detailed account of the former. It was also mentioned that if the project objectives have not fully captured the full spectrum of the objectives through

Figure 26. Bad Requirements Root Cause: Underlying Issues

the perspectives of all stakeholders involved in the project, the project requirements will most likely suffer from errors of omission.

Lastly, it was shown how faulty communication with the project stakeholders leads to missed project objectives and also to missed stakeholders, which snowballs into more missed project objectives. All of these are obviously applicable to bad requirements.

The next underlying issue for bad requirements is missing stakeholders. This is probably one of the most obvious underlying issues to understand. If stakeholders are missing and are not being included in the process of requirements elicitation, these stakeholders will most likely surface at very late stages in the project life cycle with their infamous "How come the system doesn't implement features A, B, and C? They are fundamental for the success of this project!" This will most certainly lead to cost and schedule overruns.

The next one down the ring is poor communication with stakeholders, which is an obvious direct underlying issue for bad requirements. As it was presented earlier in this chapter, T&M projects typically involve a multitude of stakeholders from several different backgrounds. The reader can probably relate to the difficulty of communication of technical subjects to a nontechnical person. This difficulty is not only valid for engineering-related technical subjects, but also for any interchange of ideas between a subject matter expert and a person who is not a subject matter expert.

The requirements-gathering process is really an exchange between project team and stakeholders. Stakeholders will present their care-abouts to the project team so they are captured as part of the project requirements. On the flip side, the project team will present to the stakeholders how the overall project requirements have been captured.

Verbiage on a typical requirements document, if that is the communication vehicle for the requirements validation, can be interpreted in different ways, depending on the background of the reader.

The IEEE standards for requirements state that requirements should be, at a minimum, unambiguous, traceable, unique, identifiable, and verifiable. However, what this book is claiming is that sometimes even if the requirements are in line with the IEEE standards, they may be misinterpreted by some stakeholders. These stakeholders may think that what they have in mind is what the requirements mean, while the project team may think it is something else.

The situation above will also lead to the infamous "This is not what I told you the system should do!" by a given stakeholder at deployment time. This will obviously lead to either a system that doesn't capture the full business value to the organization or to cost and schedule overruns to fix it.

The "users don't know what they want" issue is practically a corollary of the poor communication with stakeholders underlying issue. As mentioned above, even if requirements were captured according to the IEEE standards, there is a human aspect to it that no text can capture: how these requirements will be interpreted by a reader.

As it was said, depending on the reader's background, the interpretation of a given requirement can take many shapes. I am sure the reader, at some point in a project, has come across a situation where what is being presented to a stakeholder follows the requirements document to the letter, and yet, the stakeholder seemed surprised with what she saw.

This can be perceived as the user, or stakeholder, changing her mind as to what the system ought to implement. What this book claims is that for the most part, the stakeholders know the important things a system needs to implement. What happens at times is an interpretation problem or a translation problem between people with different backgrounds.

Since T&M projects are filled with stakeholders of multiple backgrounds, the odds of such translation problem occurring is actually fairly high. These translation problems will lead to bad requirements, even if they follow the IEEE standards, from the perspective of a system that needs to capture the full business value for the organization.

The last underlying issue is errors of omission. This, also, is almost obvious to understand in a scenario filled with many people speaking different languages, as stated above. If the project team and the stakeholders can't find a common language to fully understand each other, it is almost guaranteed that something will fall through the cracks, errors of omission. Therefore, this issue can also be related to poor communication with the stakeholders and missing stakeholders.

An important point is that this entire analysis assumes that proper project management is being applied in the execution of the T&M project. Obviously, there are a multitude of other underlying issues that can lead to project failure, if strong project management is not applied. What this book is claiming is that even in the event of strong project management under the typical frameworks currently available in the industry, some T&M projects will fail due to the reasons explained above. This book claims that these reasons have two common root causes: poor communication with stakeholders and missing stakeholders.

It is important to repeat the last sentence, as it is the golden nugget of the entire discussion presented thus far. **The two true root causes for bad requirements and poor planning, on a project where good project management has been utilized, are poor communication with stakeholders and missing stakeholders.**

With this in mind, the framework for T&M projects shall apply the people, process, and tools combo in a way that addresses these two root causes.

Also, since the utilization of system integrators by clients on the execution of T&M projects is a common practice and, as it was seen, there are several issues that can surface from this relationship, the framework for T&M projects also needs to include provisions to address these issues.

The framework for T&M projects, its people, process, and tools as well as ways to increase the odds of T&M project success when working with a system integrator are the focal points of section 2 of the book.

CHAPTER 7:
Organization Structure: The People from People, Process, and Tools

Project Team Structure

Section 1 and the section 2 introduction have set the stage for the presentation of a framework tailored for T&M projects. This chapter focuses on the people element of the people, process, and tools combo. It will cover the optimal project management structure for T&M projects that addresses the two identified root causes that drive projects to failure. It will also suggest a structure to solidify the relationship between clients and system integrators, focusing on addressing the five issues presented that can drive T&M projects to failure when they are being executed by a system integration company.

Let's kick off the discussion of the people aspect of the framework by focusing on poor communication with stakeholders and missing stakeholders, the two root causes identified in the introduction to section 2 of this book.

To recap, T&M projects usually involve a multitude of stakeholders with different backgrounds. Typically, they include stakeholders with multiple engineering backgrounds; others with finance, accounting, manufacturing, business, and marketing backgrounds; and others with no college background, but who should definitely be included as part of requirements elicitation, risk identification, and risk analysis exercises.

Typically, organizations that execute T&M projects assign a project manager to direct the project team members, collect requirements from stakeholders, perform risk identification and analysis, and be the main point of contact to all project stakeholders.

As seen previously, optimal technical communication between two parties happens when either they have similar backgrounds or at least one of the two understands enough about the subject of the other's background that there are no translation mistakes in the communication.

Let's go back to our cell phone test system example of the previous section. On this project, the list of project team includes the following backgrounds:
- Test engineering
- Software engineering
- Firmware engineering
- Embedded engineering
- Mechanical engineering
- Electrical engineering
- Validation engineering
- RF engineering

The list of stakeholders involved in our anecdotal project includes the following backgrounds:
- Various engineering skills
- Quality engineering
- Manufacturing engineering
- Finance
- Accounting
- Business
- Marketing
- Various backgrounds but not necessarily college education (operators)

What this implied then is that for our project manager to be successful in connecting with all stakeholders and project team members, she will need to have some minimum level of knowledge of all the skills listed above to allow her to avoid the translation problems mentioned.

Another aspect that is worth remembering is stakeholder buy-in to the project plan. A previous chapter showed how important stakeholder buy-in to the project plan is for the success of the T&M project. It was shown how stakeholders that fully buy into the project plan tend to be more willing to help when problems arise in the project. As it was seen, the mechanism to achieve that buy-in from the project stakeholders is not through the sale of the project plan to them, but their involvement as part of the creation of the plan, to some extent.

One aspect that is extremely important in the achievement of this level of cooperation by the project stakeholders is the overall credibility the project manager brings. The perception of competency of the project manager by the stakeholders is one of the keys to getting them involved in the project plan. If a stakeholder doesn't professionally respect the project manager, that person will most likely avoid involvement with the project at the level that we would like, ideally, as an extension of the project team. This person will probably perceive herself as not part of the team, but rather as a client of the project deliverables.

The second aspect that is a direct driver of the level of involvement from the stakeholders that we are shooting for is the project manager's personality. A person who is able to connect, on a human level, with a stakeholder will most likely get full involvement from the stakeholder in the buying-in of the project plan, or at a minimum turn that stakeholder into an advocate of the project.

In summary, the ideal project manager for our cell phone test system project should have not only skills in the technical subjects listed above, but also a personality that allows her to connect both with operators in their multiple backgrounds, all the way up to the C-level person involved in the project, passing through all the midlevel stakeholders in between.

This is an extremely tall order, borderline impossible. Nobody is this superperson alone, which is one of the reasons why a great number of T&M projects fail, especially the larger ones. It is easy to see why the larger projects are the ones that tend to fail. It is not necessarily a matter of technical complexity or bad project management, but the human aspect of it that is most times ignored.

Larger projects obviously involve a higher number of stakeholders, as they tend to have a higher impact in the organization. Also, more departments are usually touched by the larger projects, which increases the number of project stakeholders' backgrounds involved.

However, an organization can potentially get a little closer to the ideal project manager if a project management body composed of two people is formed, as opposed to a single person. Let's think through this for a second. One can potentially break down the list of skills described above into two major categories: technical skills and personality skills. This suggests that if we had a two-people combo, one being the technical brain and the other bringing the right personality, this structure would have a much better chance of success than if we spend our lifetime trying to find that one person who combines all of these traits.

Let's now take a deeper dive into each one of the two buckets, the technical and the personality. As seen in the list of technical skills for the cell phone test system project, we have engineering skills, finance and accounting, marketing, and manufacturing. One can further categorize the list above into two buckets: the engineering bucket and the business bucket. The engineering bucket would obviously include the multiple engineering disciplines. The business bucket would include finance and accounting, marketing, and manufacturing.

It would be extremely difficult to find a highly skilled engineer in several engineering subjects who would also have deep business skills. Usually, the type of personality that goes deep into one of the areas doesn't have the same level of interest for the other. It is important to keep in mind here that what we are looking for is not someone who is an expert in one and dabbles in the other. Remember that what we are striving for is credibility across the board, from all project stakeholders. We need superstars in the two areas, engineering and business.

To summarize the high-level skills we are looking for in those two people, the first one would:

- Not necessarily be the best communicator, but can get the points across
- System-level engineering expert
- Not necessarily business savvy
- Bring credibility to the engineering background stakeholders

The second person would:

- Be a great communicator
- Be comfortable with engineering subjects, but not necessarily a deep expert on them
- Be extremely business savvy
- Be a people person; have a personality that would allow her to connect in a human level with all project stakeholders, from operators to C-level

The lists suggest that we are looking for an extremely well-rounded systems engineer and for an engineer with project management experience who decided to go into the business world (maybe an MBA type) with the people-person personality.

This structure is much more reasonable from the human resources standpoint. There are people out there who meet some of the specifications we have listed. But it would be an almost impossible task to find an individual who would bring all of those traits at the level of expertise we are looking for.

With this structure, these two people would share the project management responsibility. In fact, there is a line that can be drawn as far as the responsibility that each one would have in managing the project.

The systems engineer would:
- Be responsible for the technical leadership of the project team
- Be responsible for the technical aspects of the project plan
- Be accountable to validate the estimates for the technical tasks to be executed by the project team
- Make sure the technical project team members stay on task and not fall into the traps of gold plating (engineers tend to pursue the shiny object of making things perfect; "the perfect" is definitely the enemy of "the great" in T&M projects)
- Utilize engineering expertise to earn project credibility with the engineering background stakeholders
- Lead risk analysis focused on the technical risks

The project manager would:
- Be the one responsible for the administrative aspects of the project
- Be responsible for the nontechnical aspects of the project plan

- Utilize business savvy to earn project credibility from the nonengineering background stakeholders
- Utilize personality to connect with all project stake-holders
- Lead risk analysis focused on the nontechnical risks

It is important to highlight once more that even though one of our two actors is being called the project manager, the systems engineer is as much a project manager as the other. They share the responsibility for project success. The idea is to break down the two areas of T&M project management that are somewhat mutually exclusive into two separate subroles that are part of the overall project management role for our projects of interest, test and measurements.

Another important human aspect of this change is the alignment of people with their professional interests. Nobody enjoys feeling like they are failures in their jobs, regardless of how well they are compensated by their employer. In fact, two of the top reasons why employees resign from their positions are job stress and lack of appreciation for a job well done. Both of them are actually very closely related. That is why the so-called halo effect should be avoided at all costs by organizations.

The halo effect is the idea that by putting an experienced and successful professional in a different position, the success would automatically transfer to the person's new role.

This happens often in the T&M industry. Successful engineers are "promoted" into a project management role to fix the organization's weak T&M project management. That is a fallacy. First of all, only someone extremely experienced in project management will ever be able to fix project management in an organization. Second, what ends up happening is that the organization loses a great engineer and gains a mediocre project manager.

Organizations are always much better off if they take into account the person's professional interests when making changes. If this engineer has interest in project management, she needs to be trained as a project manager, maybe take some part-time project management roles on small projects, before she is transitioned into a full-time project management role. More often than not, engineers have very little say as to whether they want to transfer into a project management role or not. Once the organization identifies a great engineer, it should invest into making that great engineer into an even greater engineer. The professional will keep her motivation level at peak, and the organization will benefit from her expanded expertise.

The approach that is being proposed here allows for these engineers to continue doing what they love, while they contribute to the management of T&M projects without being thrown to the wolves.

The Client-Integrator Liaison

This section will present a suggestion in the TMPM framework that can improve the odds of success of T&M projects that are being executed by system integration companies. To recap, a chapter from section 1 presented the top five roadblocks that can lead to project failure when system integrators are engaged:
1. Client trusts initial requirements definition to integrators
2. Client does back of napkin requirements definition
3. Over the fence mentality
4. T&M system implemented in parallel to DUT design
5. Project budget defined before feature set complete
   Let's now see how this liaison person can mitigate the five issues listed above.

1. Client trusts initial requirements definition to integrators

   As was presented in the aforementioned chapter, the
two main drivers for all the problems that surface in this
type of engagement are due to lack of time on the integrator's part in executing a proper requirements-gathering
activity and the potential lack of full picture on the integrator's part in regard to the client's target business value
for the T&M system to be built.

   Previous chapters presented the notion that project
requirements are a deeper abstraction layer that stems
from the project objectives. It was also shown that the
project objective is where the high-level care-abouts from
the project stakeholders should be captured. This higher
level of abstraction can be seen as the value proposition
statement for the T&M project. This is basically the source
of alignment between the system to be built and the organization business value for it.

   As seen in the previous section of this chapter, there
is a person in the two-people management team that is
better equipped to connect with the project stakeholders
to collect these project objectives, the person called the
project manager in the new framework.

   This is an activity that rarely is paid attention to
when a member of the system integrator staff is responsible for executing the requirements-gathering activity. The
requirements-gathering activity performed by the integrator, as it was seen, is abbreviated and usually focuses on
the test system specifications per se. But remember, garbage in, garbage out. Bad project objectives will lead to
bad project requirements, which will lead to bad system
specifications.

What is being proposed by the framework is that a person needs to perform a thorough project objective gathering, prior to the engagement of a system integrator to build the test system, even if the integrator will be given the tasks of collecting requirements.

Furthermore, there is a difference between the project requirements and test system specification. The project requirements are directly related to what the test system will do, whereas the system specification is related to how the system will do it.

What the framework proposes is that integrators are to be involved once the "what" has been thoroughly defined. They are experts in the "how," anyway. Remember the discussion around focusing people in the areas they excel at?

The liaison role that is being defined here would be the person responsible to collect the "what," which, again, will stem from the project objectives, which is the business value for the test system that would have also been collected by the same person.

2.   Client does back of napkin requirements definition

As detailed by a previous chapter, the facts listed below are usually present in this type of engagement with integrators:

—   The client may not be as well versed in performing a thorough requirements analysis for T&M system as an expert from the field would be
—   Common for internal resources not to have in-depth knowledge of the type of information an integrator usually needs from clients in order to keep headed in the right direction

— Internal resources rarely can dedicate themselves full time to the requirements-definition exercise
— It is usually extremely difficult to find resources who can execute a well-balanced business and technical requirements-gathering activity

This is almost a list of desired qualifications for the person occupying the liaison role. The end results, much like what was presented for case number one above, is that, at a minimum, the client should have ready for the integrator the comprehensive list of project objectives and the "whats" or project requirements for the test system. Some clients may want even to go one step further and define the test system specification, which is the third level of abstraction for requirements.

A challenge with this approach is that some organizations may not have the need for a full-time person executing this activity. In this case, a consultant may be the solution.

3. Over the fence mentality

This situation happens when the client believes that since an expert company is being hired to execute the T&M system, the system integrator will take care of everything with absolutely no or minimum interaction with client internal resources, until it is time to deploy the system. In this described scenario, not only the requirements-gathering activity is given to the integrator, but there is usually minimum interaction throughout the project life cycle.

This type of engagement usually happens when the client doesn't have the needed technical resources in house to manage the work performed by the integrator. This is actually a bad situation for both client and integrator.

As was seen in a previous chapter, the main issues that this type of engagement brings are:
— Lack of alignment of the final delivered system with the client's business value
— Information flow from client to integrator suffers, which may lead integrator to be headed in the wrong direction
— Unreasonable expectations on the client's part, mainly due to ignorance of what can and what can't be accomplished within the project schedule and budget

This case also can benefit from the suggested liaison role. This person would need to understand the integration business enough to make sure the needed information from the client to the integrator is flowing at opportune times. The liaison would also need to be technically competent on the T&M subjects in order to follow the technical implementation that is being proposed by the integrator. Ultimately, this person would make sure the business value collected as part of the project objectives are always kept in line with the project implementation and would keep the client's expectations in touch with reality.

4.   T&M system implemented in parallel to DUT design
      This issue calls for a deeper solution than just simply assigning a liaison to coordinate the interface between client and integrator. As it was mentioned by a previous chapter, this situation can't really be properly addressed if the project is being executed in a waterfall format, due to the nature of the fluid requirements.
      In this situation, the client is better off by setting up a partial agile structure and having the liaison function as a product owner for the agile project. In order to better un-

derstand what this entails, we should go through a brief introduction to the agile principles. The agile subject is beyond the scope of this book; however, there are several good books on the topic if this interests the reader. For the purposes of this book, only an extremely brief introduction on agile will be provided.

The Agile Methodology

In 1970, Dr. Winston Royce presented a paper entitled "Managing the Development of Large Software Systems," which criticized sequential development. He asserted that software should not be developed like an automobile on an assembly line, where each piece is added in sequential phases. In such sequential projects, every phase must be completed before the next phase can begin. Dr. Royce recommended against the phase-based approach in which developers first gather all of a project's requirements, then complete all of its architecture and design, then write all of the code, and so on.

The waterfall methodology assumes that every requirement of the project can be identified before any design or implementation occurs. This clearly goes against the very foundation of an NPI process where the T&M project needs to start some time before the unit under test development is completed, making it impossible for the project requirements to be defined beforehand.

Agile development methodology provides opportunities to assess the direction of a project throughout the development life cycle. This is achieved through regular cadences of work, known as sprints or iterations, at the end of which teams must present a potentially shippable product increment. By focusing on the repetition of abbreviated work cycles as well as the functional product they yield, agile methodology is described as "iterative" and "incremental." In waterfall, development teams only have one chance to get each aspect of a project right. In an agile paradigm, every aspect of development—requirements, design, etc.—is continually revisited throughout the life cycle. When a team stops and reevaluates the direction of a project every two weeks or any other short interval that makes sense in the context of the project, there's always time to steer it in another direction.

This "inspect-and-adapt" approach to development is a perfect match to the fluidity of the NPI cycle. Once the requirements to be implemented by each sprint are aligned with the incremental progress that is made in the parallel development of the device under test itself, progress can be made on the T&M project on areas that have a lower risk of being changed due to a change in the device under test being developed. Since each sprint is of short duration, it gives stakeholders recurring opportunities to calibrate releases for alignment with changes in the device under test development. This allows for a much leaner implementation of the T&M system, reduction of waste and rework due to unit under test changes, and consequently maximization of the business value for the test system.

Scrum is the most popular way of applying agile methodology on a project. The responsibilities of the traditional project manager role are split up among roles (any similarity with what the TMPM framework proposes for its organizational structure is not mere coincidence). The main roles defined under Scrum are the product owner, the Scrum Master, and the project team.

The project team takes a much more active role as far as accountability for the project success. The nature of Scrum is that the teams are self-organizing. What that means is that since the duration of a sprint is short, the number of requirements to be implemented is much lower than the total number of requirements for the entire project. With that, each team member has better visibility of where the finish line for that sprint is, which allows them to make more independent decisions as to the tasks that need to be accomplished and their priorities.

The Scrum Master is a facilitator. Her main task is to remove the impediments to the ability of the team to complete the sprint deliverables by the end of the sprint.

The role of highest interest for the discussion of this section is the product owner. The product owner represents the voice of the customer, or in our T&M project case, the voice of the stakeholders. Her main task it to ensure that the value to the business is being delivered by the project. The product owner is the one that writes what is called "user stories," which can be seen as features to be implemented; ranks and prioritizes them; and adds them to the "product backlog."

The product backlog is an order list of requirements that is maintained for the project. This can be seen as the overall project requirements. The main idea behind Scrum is that a number of these items from the product backlog, the ones with higher priority numbers, are selected to be worked on by the project team in the course of the next sprint. Any requirement that doesn't make the cut for the next sprint is seen as an out-of-scope item and is not touched at all.

Now, going back to our NPI scenario, one can infer that a structure like the one mentioned above can be extremely advantageous to the project. The initial requirements list—or, to use the agile terminology, the product backlog—can assume the test system will be built by taking the current state of the device under test as if it has been already completed. The product owner would then prioritize and rank the product backlog in a way that the stable features would be addressed first.

At the end of each sprint, the product owner would audit the current state of the device under test, speak with the product development stakeholders about any possible upcoming changes, and collect the overall direction of the product development effort. This information would then be used as an input to another exercise in reprioritizing/reranking the product backlog remaining items for the requirements of the next sprint to be formed.

This practice may not fully eliminate rework altogether since, depending on the fluidity of the particular NPI cycle, there still may be cases where the T&M project team is asked to take a step back and rework functionality that addressed what was initially seen as stable features of the new product. However, this will certainly increase the odds of minimizing these step backs along the life of the project. This result not only increases the odds of overall project success around business value, cost, and schedule, but also brings another very important by-product. It reduces the overall frustration of all people involved in the project:

- The stakeholders will see tangible progress being made, sprint after sprint, toward the final test system to be delivered
- The project team members will probably not come across the "users don't know what they want" syndrome as often
- The initial project budget and schedule, if defined with the agile mentality, is probably maintained to a much closer level of accuracy than in the usual waterfall method of implementation for both client and integrator

With this in mind, the liaison person this book advocates for improving the client-integrator relationship would be the perfect candidate to function as the product owner for this implementation. This person would:

- Be the one collecting the initial list for the product backlog (more on this in the next chapter)
- Working with the project stakeholders to prioritize and rank the product backlog items
- Determine the breakdown of the scope for each sprint as well as the estimated number of sprints
- Coordinate the procurement process with the integrators, by directing them to quote the project in a way that there would be pricing allocated for each of the sprints separately
- Work with the client to secure initial overall project budget for the T&M effort in a way that would leave management reserve beyond the sum of all sprint prices in order to account for changes driven by the NPI process that can't be accommodated by the reprioritization of the product backlog alone; e.g., new instruments that need to be added due to a new device under test feature that made the NPI cut and now needs to be tested
- Functions in the agile definition for the product owner during the course of the T&M project implementation

As it was mentioned previously in this book, integrators and clients are not used to working together under this format. Clients expect that integrators will honor the original project budget while delivering maximum business value. Integrators try to defend the original project scope as if their lives depended on it, which, in fact they do, if there is no wiggle room for change. The end result is overall frustration and project failure.

The next chapter will present more details on the mechanics of implementing a modified agile structure for T&M projects in support of NPIs.

5.   Project budget defined before feature set complete

As presented in a previous chapter, it is common for clients that defining the business need for a T&M system to need some dollar figure to justify that capital expense. Usually, a procurement process is started and integrators are invited to provide technical proposals for implementation. Very often at that early stage, the full feature set for what the T&M system will implement is not necessarily flushed out. It is not uncommon for clients to request a ROM (rough order of magnitude) estimate for capital budget to be secured.

This is very similar to the previous scenario of a T&M system supporting an NPI. In this case, the project objectives and requirements set have also not been defined, there is a great chance of changes down the road once the stakeholders become more familiar with the project, and integrators are asked to come up with pricing for the overall project under those circumstances.

What this book suggests is that client and integrators would be much better off if the modified agile methodology is applied, as described in the previous scenario. In this engagement type, the role of the liaison would be basically as described above.

To summarize the liaison role, the lists below describe the responsibilities for this role and the experience level a liaison candidate would need to have.

<u>Role Responsibilities</u>

- Execute a well-balanced business and technical collection of project objectives
- Execute a well-balanced business and technical requirements-gathering activity

- Execute a well-balanced business and technical risk analysis
- Ensure alignment of the final delivered system by the integrator with the client's business value through the monitoring of project progress throughout the entire project life cycle
- Make sure information flows from client to integrator in a timely and accurate manner throughout the entire project life cycle
- Make sure to align the client's expectations on what is possible and what is not, based on the project parameters for cost, schedule, and scope

In a modified agile approach:

- Be the one collecting the initial list for the product backlog (more on this on the next chapter)
- Working with the project stakeholders to prioritize and rank the product backlog items
- Determine the breakdown of the scope for each sprint as well as the estimated number of sprints
- Coordinate the procurement process with the integrators, by directing them to quote the project in a way that there would be pricing allocated for each of the sprints separately
- Work with the client to secure initial overall project budget for the T&M effort in a way that would leave management reserve beyond the sum of all sprint prices. This would be so changes driven by the NPI process that can't be accommodated by the reprioritization of the product backlog alone are accounted for in the beginning; e.g., new instruments that need to be added due to a new device under test feature that made the NPI cut and now needs to be tested.

- Functions in the agile-defined product owner role during the course of the T&M project implementation

Needed Skills and Personality

- Be a people person to best perform collection of project objectives, project requirements, and initial project risks
- Business savvy to make sure the business objectives and requirements are properly captured
- Expert in requirements gathering for T&M projects
- Expert in risk analysis for T&M projects
- Deep knowledge of the system integration business in order to understand what type of information needs to be conveyed to the integrator as well as timing of that information flow
- Skilled on technical aspects of T&M projects in order to follow the proposed solutions by the integrator as well as make sure the project is progressing toward what the final solution ought to look like

In a modified agile approach:

- All of the skills and personality listed above
- Experienced in agile implementations
- Experienced in fixed-price contracts in agile format

This concludes the organization structure for the proposed TMPM framework. This chapter showed the benefits of a two-people project management body instead of a traditional approach with a single T&M project manager. It also proposed a new role to increase the odds of success when a system integrator company is put in charge of implementing the T&M project.

Lastly, it introduced the concept of an agile-based methodology for the execution of T&M projects when the test system is to either support an NPI or when the project budget needs to be defined prior to the test system feature set being fully defined. The next chapter will focus on the tools element of the people, process, and tools combo for the proposed new T&M framework.

CHAPTER 8:
System Modeling: The Tools from People, Process, and Tools

The previous chapter introduced an organization structure for the management of T&M projects that allows the two people responsible for managing the project to be better aligned with their skills, areas of expertise, and interests. Also, it better addresses the stakeholder management problem on T&M projects, which notoriously includes many stakeholders with different backgrounds.

It also proposed the creation of a role to liaise the relationship between client and system integrators when one is involved in building the test system. It was shown how this role specifically addresses the issues usually present in the engagement of integrators.

Lastly, it introduced the concept of how to apply agile methodology in projects that need to support NPIs and ones where the project budget is defined prior to the system feature set being complete.

This book will purposely postpone the process discussion until the next chapter and will focus this chapter on the tools element of the people, process, and tools. Since the TMPM process leverages this tool set very heavily, it will allow the reader to better comprehend the process component once the tools have been presented in detail.

Introduction to UML

Before jumping into the tools discussion, we need to baseline our knowledge on a subject that is the basis for the tool set, UML. Though a comprehensive account of UML is beyond the scope of this book, this author strongly suggests that the reader who is interested in the process proposed herein acquire a book on UML that is available in the literature. There is a mention or two of those in the references section of this book. This text will focus on specific areas of UML that will serve as the foundation for the tools and process discussions.

The Universal Markup Language, UML, is a general purpose modeling language. It was introduced in 1997 by the Object Management Group (OMG) with the original intention of providing information technology professionals with a stable and common language that could be used to communicate software designs among multiple software developers.

UML is a set of different types of diagrams, each of which can be used to represent a level of abstraction, or level of details, of a system. One can use a specific diagram type to model the highest level of a system, and drill down into the details of each of the high-level components by utilizing other types of UML diagrams that are better suited to represent different levels of detail. It offers a standard way for people to visualize the system blueprints.

UML version 2.2 has fourteen different types of diagrams that are available for use, broken down into two high-level types: structure diagrams and behavior diagrams.

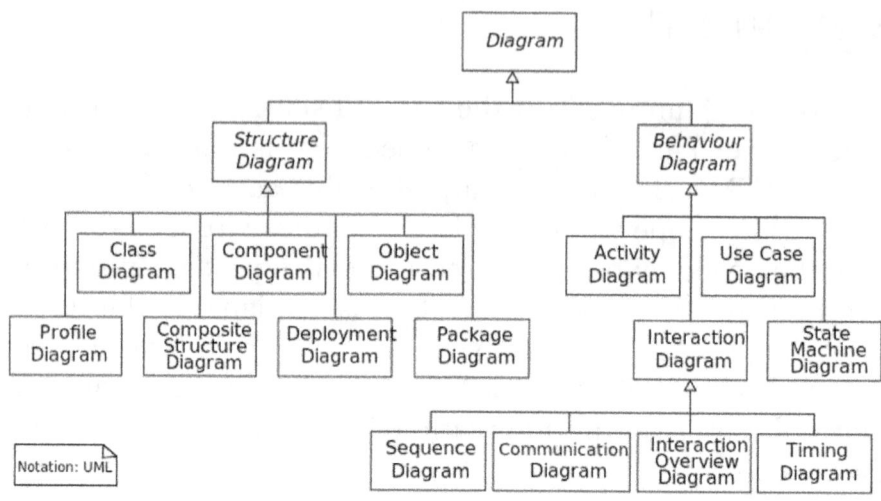

Figure 27. UML 2.2 Fourteen Diagrams

Behavior diagrams emphasize what must happen in the system being modeled, or the behavior of the system, or, why not say, the functionality (requirements) of the system. One can intuit that these types of diagrams would be good candidates to visually represent "what" the system must do. One may remember that the project "whats" are related to the project objectives and project requirements.

Structure diagrams emphasize the things that must be present in the system being modeled, or "how" the system will implement "what" the system needs to do. One can also intuit that these types of diagrams would be great candidates to model the system design, the "hows" of the system being built.

At its core, one of the strongest points of UML is that it provides a visual representation of a set of functions in a nontechnical manner. As we will see below, the diagrams present a visual language that is very intuitive for one to follow, and at the same time provides standardization, one of the features of any language.

Another clear advantage of UML is that a system can be fully modeled either by taking advantage of all fourteen different UML diagrams, or it can be fully represented through the utilization of just a subset of those diagrams. Simpler systems, with less interconnectivity between departments and with less overall stakeholders, will yield a simpler overall UML representation than a very large complex system. We will skip the detailed explanation of each of the fourteen diagrams and will focus on the ones that are directly used as a basis for the TMPM framework. Also, we will introduce each diagram at opportune times in the book, as they are needed as the foundation for the topic under discussion.

For the time being, let's focus our attention to two behavior diagrams, use case diagram and activity diagram.

Use case diagrams describe the functionality provided by a system in terms of actors, their goals represented as use cases, and any dependencies among those use cases. The actors represent elements that will interface/utilize the system, either human beings or other systems that are external to the system being modeled. Figure 28 shows a use case example that models the functionality of a portable audio player. One will notice that this system will have three main high-level functions: operate audio player, maintain playlist, and maintain audio player, all of them to be performed by the "listener" actor.

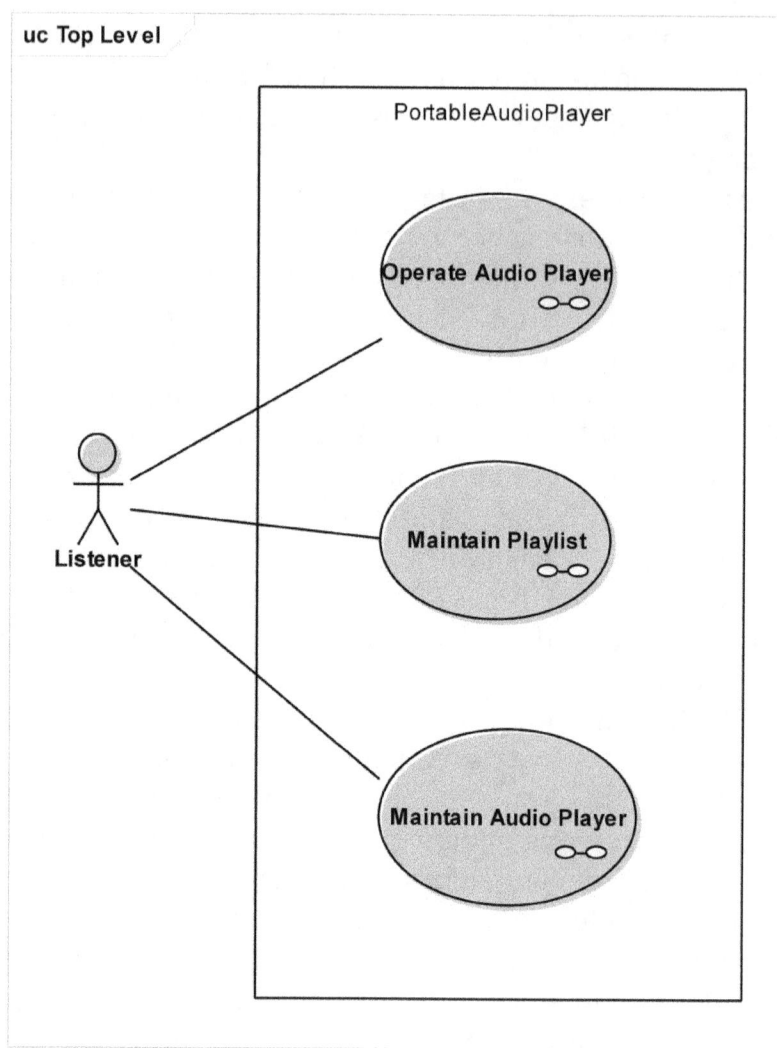

Figure 28. Example of Use Case Diagram

The next natural activity to model this system would be to drill down and define the use cases that are involved in each one of those three high-level functions. Figure 29 shows the use case diagram for the "operate audio player" use case.

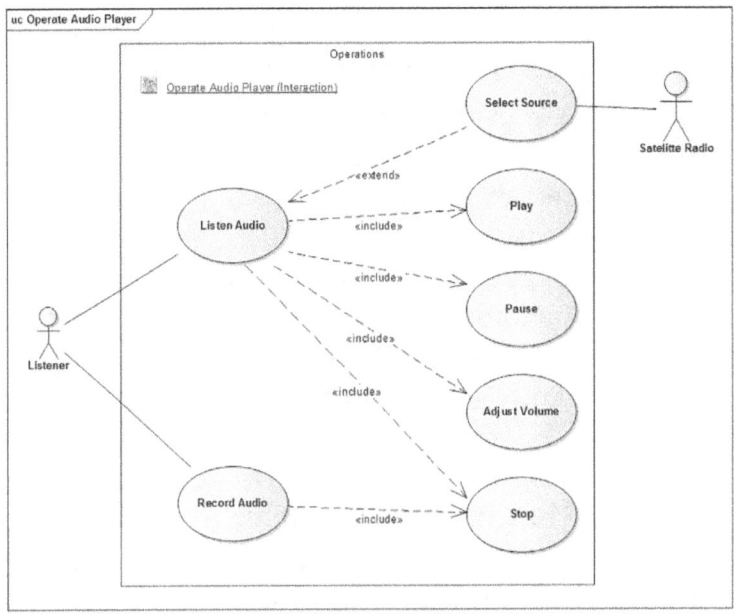

Figure 29. Operate Audio Player Use Case

If the modeler would like to divide higher-level use cases into a few more detailed ones, she should use one of the two available use case relations: include and extend. Include is used to separate a logical sequence of actions from the higher-level use case. The extend relation models optional actions from the higher-level use case.

The way to interpret what is being modeled above is that there are two high-level functions on the "operate audio player" use case, listen audio and record audio. When the listener actor executes the record audio function, the record audio includes the action of being stopped, which will need to happen at some point. When the listener actor selects listen audio, it will be played, paused, the volume adjusted, and stopped. Therefore, all of these other use cases are "included" in the two high-level use cases.

Another way to look at the include relationship is that the lower-level use cases can potentially be "included" by multiple higher-level use cases as part of its functionality. For example, in the diagram above, the stop use case is used by both listen audio and record audio use cases.

The last use case to be explained is "select source." Notice that when the listener is executing listen audio, it has the option to select the audio source from MP3 list or satellite radio. However, the user doesn't necessarily need to do that, as it will have a default selection. Therefore, this use case has an "extend" relationship, or, in other words, optional relationship with the listen audio use case. Also, notice how the satellite radio actor has been represented. It is located outside the boundary box, as it is an external element to the system being modeled, but still may be accessed by the select source use case. The MP3 list, the other source the player can select, hasn't been represented as an actor because it is maintained within the scope of the audio player.

The use case diagram is usually one of the first diagrams that is created in order for a system to be modeled.

The next UML behavior diagram to be explained is the activity diagram. Figure 30 shows an activity diagram example that models the select source use case on its multiple activities.

One can rightly infer that each one of the use cases described in a system with multiple use case diagrams will most likely have an activity diagram detailing its functionality. It is also correct to interpret the activity diagram as the next abstraction layer from the use case diagram. It shows the procedural flow of control while processing an activity. It is probably one of the most flexible UML diagrams as it can be used to model high-level business objectives and project requirements as well as low-level implementation relationships.

Another advantage of this type of diagram is that it describes not only the flow of activities in a sequential fashion, but also shows who is responsible for each of the activities.

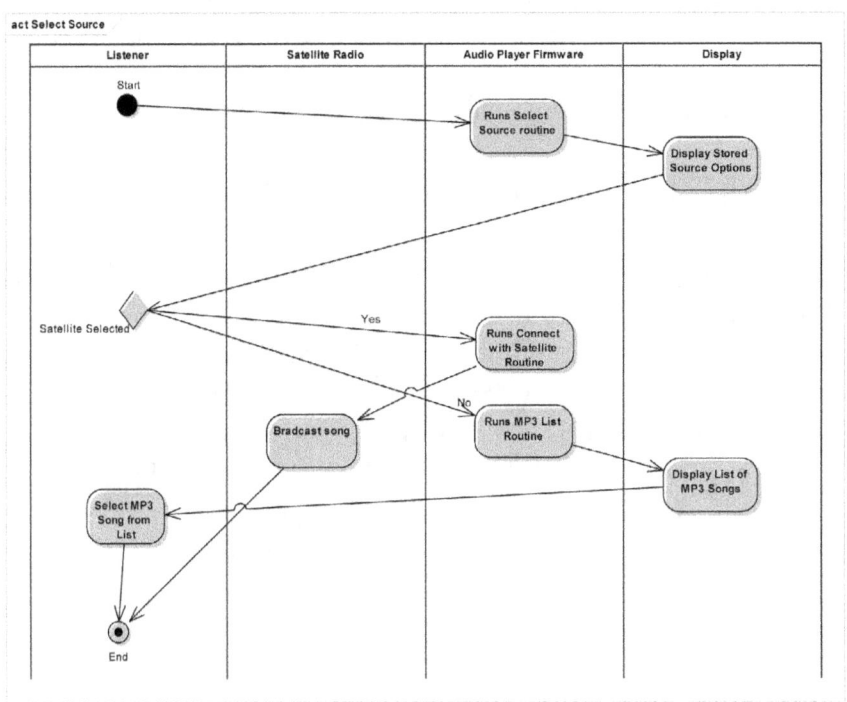

Figure 30. Select Source Activity Diagram

The diagram above shows how the select source use case activity would unfold. In this implementation of the activity diagram, it mixes high-level information on who performs the selection and which actors are involved in the activity as well as low-level information on which firmware routines are invoked and what shows up in the device display as a result of those routines executing.

Depending on the goal of the activity diagram, the information captured by the diagram may vary. One may elect to keep the relationships in a higher level if utilizing the diagram to collect business objectives and project requirements, or in a lower level if describing system specifications or even design.

Modeling Project Objectives and Requirements

As presented previously in this text, one of the problems that needs solving in T&M projects is poor communication with stakeholders, which have corrosive effects on the foundation of T&M projects. In these types of projects, there are an increased number of stakeholders with differing backgrounds. Therefore, a methodology for communication with these stakeholders needs to take into account this very fact.

The typical methodology used for collection of project objectives and requirements has been interviews with the project stakeholders and the collection of the information in some sort of text document. The verbiage presented in this document should capture the nature of the objectives and requirements. Moreover, the text should be written in a way that it represents accurately what the stakeholders intended as well as what the requirements-gatherer person interpreted, and the two aspects need to match in order for the requirements to be valid.

The practice usually consists of the requirement gatherer asking questions of the stakeholders, which then provide answers to those questions that get captured in the project objective/requirements document. If the process is thorough, as it needs to be, several stakeholders from all areas the project will touch are interviewed, and their answers are captured in the document.

There are several problems with this approach. The first is the fact that the information gathering in this format will have a question-and-answer format. The person being inter-viewed will basically answer whatever question is being asked and wait for the next question. The onus of describing what the system is expected to do by asking the right questions will fall solely in the shoulders of the interviewer. The questions better cover the entire spectrum of scope the system ought to implement; otherwise, there will be objectives/requirements that won't be gathered. It is not to say that the person being interviewed is sandbagging the process; it is flat-out human nature. Once a question-and-answer interview format is estab-lished, the interviewee will most likely just answer questions, as opposed to helping the interviewer paint the whole picture of the project objectives and requirements.

A better approach to the interviews would be a conver-sation where some sort of picture is being painted, literally, while the conversation takes place. Think back to the UML behavior diagrams that were presented in the previous section. Think of the interviewer asking some questions and building the use case and activity diagrams as the stakeholder answers those questions. At that point, it becomes a fill-in-the-blanks exercise instead of a question-and-answer one. The task now, from the interviewee perspective, is not just to answer ques-tions, but to complete the diagrams that are being drawn. This is much more conducive to unsolicited information being vol-unteered by the stakeholder, even if there isn't a specific ques-tion being asked targeting that information.

It becomes much easier for the stakeholder to get the ideas flowing out of her head and materializing onto the dia-gram. That is exactly what we want. We want to make sure a literal brain dump is captured from the stakeholder's mind. The modeling process is much closer to a brain dump than the usual question-and-answer, write-answers-down-in-a-text-document methodology.

Another issue with the typical approach is the fact that it is extremely difficult for the requirements gatherer to think, or even know a priori, all branches of questions that need to be asked during the interview. As mentioned before in this text, the stakeholders are the ones in the best position to know what the system is supposed to look like in its entirety and all aspects of their jobs that will be influenced by the system to be built. Even if the requirements gatherer is extremely experienced in building test systems, it can't possibly think of all areas of a stakeholder's job that can be touched by the system. This leads to some missing questions, and the consequence, as seen above, will be missing requirements in the question-and-answer requirements-gathering format.

As was mentioned, the requirements-modeling approach through a visual representation allows the stakeholder to take the driver's seat and direct the branches that need modeling. The requirements gatherer will be merely a facilitator at that point, which is the best position she can possibly be in. The star of this process is not the person gathering the objectives/requirements, but the stakeholders.

Another easy to visualize the advantages of the modeling methodology versus the text document approach is the identification of exception cases, the infamous "what if" scenarios. Think of a requirements-elicitation process where the requirements gatherer is typing up a text document as questions get answered. It becomes extremely difficult for the group performing the exercise to think of asking a "what if" question.

For example, imagine a requirement-gathering exercise where the requirements-gathering person asked the stakeholder, "What happens when the breakout board battery is connected to pins one, two, and three of connector J10 to simulate a battery?"

It is easy to see an answer of the type, "The system will need to measure a voltage at connector J8 pin twenty-four." Mission accomplished in this line of questioning, right? The interview will probably move on to other aspects of the system to be built. Now, instead, think of a requirements-modeling exercise where the following is captured:

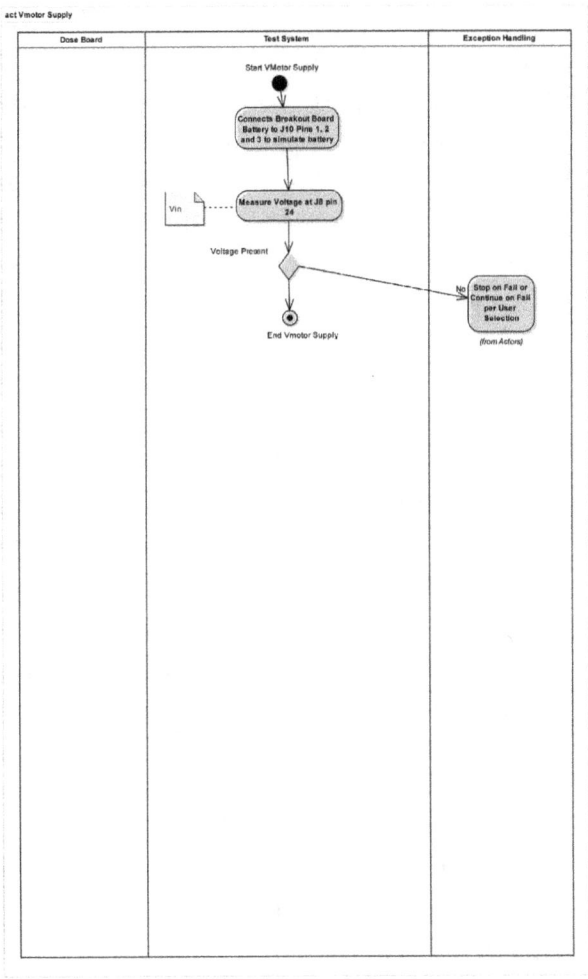

Figure 31. "What If" Activity Diagram

The activity diagram above presents the exact same information as the question and answer that was captured in the question-and-answer approach to collecting requirements. However, one can see that a "what if" question is almost begging to be asked when one looks at the diagram. What if a voltage *can't* be measured at J8 pin twenty-four?

In that case, the system needs to handle the exception somehow. "How?" the requirements gatherer would ask. "Well," the stakeholder would say, "in that case, the system shall pop up a question for the operator that would select either continue or stop if fail." As one can see, modeling the requirements allows for exception cases to be easily spotted and a discussion to be had on possible ways to handle them.

One last issue with the original text-based approach to collecting requirements arises at review time. As previously mentioned, ideally, the project team needs to always strive for buy-in by the stakeholders. It was also shown that the best approach to getting buy-in is by involving the stakeholders as part of the solution as opposed to dumping the solution on them and asking for approval.

Now imagine the requirements document of a very large and complex T&M project. It will most likely include hundreds of pages of text. If the requirements buy-in mechanism consists of an e-mail with the requirements document attached asking for a review, the stakeholder's only mechanism to reviewing the document is to read through it, on his own. One of the goals of a requirements review is for stakeholders to catch mistakes and errors of omission in the requirements.

What do you believe the chances are of mistakes and errors of omission to be spotted by a stakeholder review on a two-hundred-page requirements document? Even if the reviewer is well intentioned in adding value during the review, by the time the fifteenth page of requirements rolls in, she will start thinking about her son's soccer practice in the afternoon, that she needs to pay her credit card bill today, and that tomorrow night is date night with her husband.

This type of review is not very conducive of the goals of catching mistakes and errors of omission; it is more of a checkbox in a process that needs to be checked off. Now, instead of this type of review, imagine a meeting with several stakeholders, led by the requirements-gatherer person where the full set of requirements-model diagrams is the focus of review. This methodology has a much higher chance of keeping the stakeholders engaged and having them really taking ownership of the project objectives/requirements.

Modeling System Specification and Preliminary Design

The next step down in the project life cycle is the determination of the T&M system specification and the overall system design. These two are obviously separate aspects of a T&M project; however, in reality, they usually navigate very close to each other.

The system specification is the detailed function and technical description of a system. Though they are still related to the system "whats," they actually usually start taking into consideration the "hows." The system specification can be seen as a bridge between the project requirements and the high-level system design. It further details key performance parameters that are dictated by the project requirements as well as defines the key system attributes, which are usually the first touch points to the high-level system design. Or, in other words, it derives the system requirements to a preliminary design level of details.

Usually, the system specification on a T&M project is the document that allows refining of cost and schedule as well as the establishment of project baselines for its metrics. It also ensures traceability of the project objectives and requirements to a system design, making sure the selected design approach is taking into account all system requirements. It is the link from the "how" back to the "what."

Previously in this chapter, it was mentioned that the UML structure diagrams emphasize the things that must be present in the system being modeled, or "how" the system will implement "what" the system needs to do. One can also intuit that these types of diagrams would be great candidates to model the system design, the "hows" of the system being built.

Since we are now talking about the system design, let's introduce three UML diagrams that are extremely useful in the exercise of modeling the system design: class diagram, component diagram, and deployment diagram.

The class diagram is a type of static structure diagram that describes the structure of a system by showing the system's classes or the things that comprise the system. Though a class is a construct of object orientation for software programming, it is important to highlight that a class can very easily take other forms of constructs. It can represent a logical block of a system or even a hardware component. The main idea that needs to be understood is that a class can be anything that can be self-contained in a module of sorts, the building blocks of a system.

The class defines the attributes and behaviors that a system component will implement. The behavior is described by how it reacts to interfaces of other classes as well as other functions that it executes. The class diagram therefore shows how the system is being broken down into modules, as well as the static relationships between these modules.

The diagram below illustrates aggregation relationships between classes. The lighter aggregation indicates that the class Account uses AddressBook but does not necessarily contain an instance of it. The strong, composite aggregations by the other connectors indicate ownership or containment of the source classes by the target classes; for example, Contact and ContactGroup values will be contained in AddressBook.

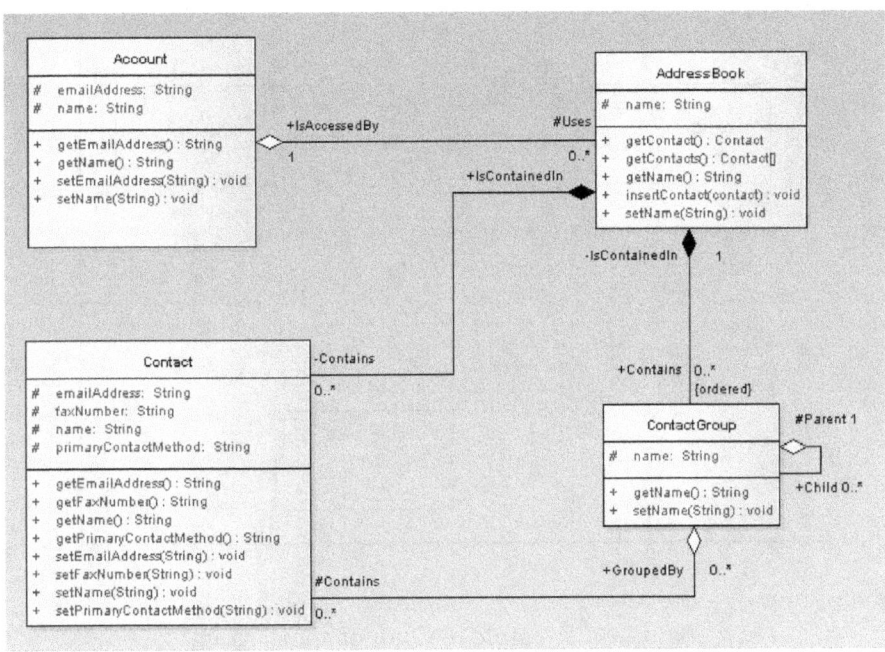

Figure 32. Example of Class Diagram

The next immediate higher level of abstraction to the class diagram is what is called the component diagram. A component is implemented by one or more classes into larger logical building blocks. Components are great to compartmentalize classes into logical blocks of functionality. Again, the same comment that was made for classes is valid for components. Even though they started off as constructs to be used on modeling of software systems, they can very well represent hardware logic blocks, such as controllers, instruments, fixtures, custom electronic boards, etc.

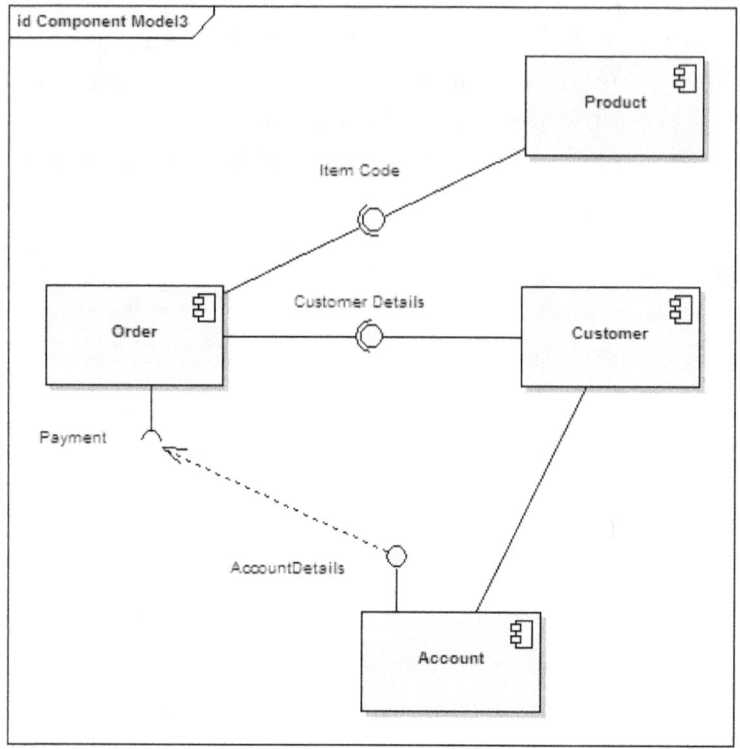

Figure 33. Example of Component Diagram

Figure 33 demonstrates some components and their in-
terrelationships. Assembly connectors "link" the provided
interfaces supplied by Product and Customer to the required
interfaces specified by Order. A dependency relationship
maps a customer's associated account details to the required
interface, Payment, indicated by Order. Components define
boundaries and are used to group elements into logical struc-
tures. The component modeled above sets a very clear boundary that only classes Order, Account, Customer, and Product
are part of it.

The last diagram that is worth mentioning in this discussion is the so-called deployment diagram. A deployment diagram models the run-time architecture of a system or, in other words, the running test system. It shows the configuration of the hardware elements (nodes) and shows how software elements and artifacts are mapped onto those nodes.

Figure 34 shows an example of a deployment diagram.

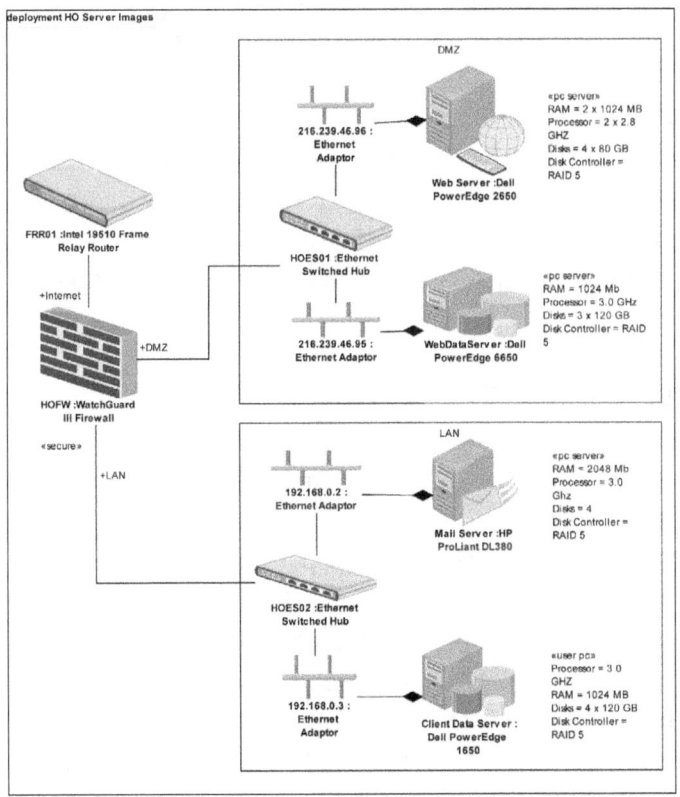

Figure 34. Example of a Deployment Diagram

As one can see in the example diagram above, deployment diagrams basically detail all different deployment nodes that will compose the system and the communication path between nodes.

Multiple levels of abstraction can be used on deployment diagrams also. For instance, the example above may be used to communicate high-level deployment to a group of stakeholders. Maybe, for a different group of stakeholders, who may care more about what runs on each of those nodes, lower-level diagrams can be created and associated to each node. These diagrams would show what components are being deployed to each node.

It is not difficult for one to see how these three UML diagrams, when used together, can not only fully represent the system specification and high-level design, but also represent the full system design when the appropriate levels of abstraction are used.

Stakeholder Management

Now that tools for both requirements and design modeling have been presented, the next important tool to be applied by TMPM is one to manage the project stakeholders. As seen in a previous chapter, the tailored T&M framework ought to address two main root causes that are the source of all issues that can surface on T&M projects: poor communication with stakeholders and missing stakeholders.

The modeling tools presented above certainly address the communication with stakeholder issue. They also indirectly address the missing stakeholders issue, since, as was seen, the requirements-elicitation process is an iterative process. As the stakeholders get more and more familiar with the system requirements, they may identify other

stakeholders that should have input on the modeling process as well, ones that may not have been initially identified
as project stakeholders.

To stitch this together, we need a mechanism to manage
these stakeholders. The first task is to create what is called
the stakeholder register, which is basically a list showing all
project stakeholders. Table 41 shows an example on how to
capture and maintain this information.

Project Stakeholder Register

Stake-holder Name	Project Role	Organi-zation	Contact Infor-mation	Main Require-ments	Interest (1–5)	Influ-ence (1–5)

Table 41. Project Stakeholder Register

This table contains information that will be used to create
the stakeholder management strategy, which will be detailed a
little later in this section. We shall now describe each entry of
the table and explain how the information will be utilized in
creating the stakeholder management strategy.

Each stakeholder will have a role in the project. It is important to think of roles not necessarily being of active participants of the project team, but being basically anyone who
will interact with the project in any way. Referring back to the
cell phone test system project example from the last section,
the roles that were defined as the list of stakeholders for that
project would basically be transcribed into table 41.

The organization column shows the organization division if internal to the organization or the name of the corresponding organization if external to it. The contact information will include the stakeholder contact information and will highlight the preferred method of contact.

The main requirements column provides a high-level description of what each stakeholder is expecting the project to provide. Needless to say, this information is extremely important for the project manager to align all expectations. Gross misalignments need to be dealt with as soon as possible in the project life cycle. Either the stakeholder needs to be aware that what he is expecting is out of the scope of the current project or the project scope needs to be reviewed and potentially changed if the stakeholder's expectations were somehow missed during the process of creating the business case and the project charter. Always keep in mind that the sooner problems are identified, the cheaper they are to fix. Imagine an important expectation from an influential stakeholder that is somehow missed through the project life cycle and only gets identified at the Closing process group. Depending on the missed expectation, the entire project might be scrapped. Make sure to enter no more than five bullets in this column for each stakeholder. At this stage, it is important to capture only the high-level expectations in order for gross misalignments to be spotted.

The interest column shows how important this project is for each stakeholder. A stakeholder that is only filling in part time in a small project role will have a different level of interest than a person whose job depends on the success of the project. Also, it is important to gauge the level of interest of members of the executive management, the C-level stakeholders. This level of interest is somewhat a gut feel, but a savvy project manager can gauge the level of interest of executives by observing their level of involvement and how the project aligns with their professional goals within the organization. This is more of an art than a science, but it is valuable information to capture nonetheless. It will be used as an important input to the creation of the stakeholder management strategy.

The next column, influence, is self-explanatory. Stakeholders who have veto power automatically jump to the top of the scale, while others who don't have that much power to affect the project will lean toward the other end of the spectrum. This information might seem unimportant at this point, but it will serve its purpose in the creation of the stakeholder management strategy.

The stakeholder register is expected to be a living document. While the project unfolds, it is important to always be vigilant for stakeholders that might not have been identified. After all, missing stakeholders are one of the root causes that we are trying to address with the TMPM framework. Any new stakeholder identified along the course of the project needs to be captured into the stakeholder register and the proper high-level analysis need to be performed in order for the information to be filled out. Good project management is all about thinking ahead and anticipating problems as opposed to taking action. Be aware of project managers that are always appearing as heroes and saving their projects from disaster. They are heavily relying on luck while managing their projects as they assume they will always be able to handle problems once they arise. Their luck usually runs out at the worse possible moment and catastrophe strikes. The really valuable project managers are the ones that seem not to work that hard on their projects while their projects are usually uneventful. They give the impression that their projects are always lucky to not come across any unforeseeable event. The truth of the matter, though, is that they are probably spending the majority of their time trying to predict what can go wrong and steer their project away from the rough waters.

The stakeholder management strategy will use the information collected during the process of creating the stakeholder register, and derive a management strategy for each one of the registered stakeholders.

It is important to keep in mind, as was emphasized, that the best project managers focus mostly in thinking ahead before taking action. Having a documented management strategy in place for all parties involved will help the project manager make decisions along the course of the project that are in line with the identified management strategy, as opposed to not considering the human aspect of things at all. Usually, decisions that are made taking into consideration the stakeholder management plan are the ones that can keep the project moving in calm waters as opposed to potentially driving it through conflicting interests of the multiple stakeholders.

The main goal for the stakeholder management strategy is to outline a plan to increase support and minimize obstruction from the stakeholders, **before crisis strikes**.

Stakeholders can be managed individually, or they can be grouped together so management strategy is devised on a group basis. The later is especially needed in the case of hundreds of stakeholders; however, they tend not to be needed for typical T&M projects. For example, stakeholders can be grouped geographically, by level of interest, by level of influence, or by any other way that makes special sense in the project at hand.

Another important data point that should be captured is the expected level of participation desired by each stakeholder, that is, how much information they would like to receive, how much involvement they would like to have in the multiple phases of the project, etc.

You may remember that two pieces of information gathered as part of the stakeholder register were the levels of influence and interest. Stakeholders can be classified as being part of one of the following four quadrants: Quadrant 1, High Influence/Low Interest; Quadrant 2, High Influence/High Interest; Quadrant 3, Low Influence/High Interest and Quadrant 4, Low Influence/Low Interest.

As mentioned before, imagine a project with hundreds of stakeholders registered. A stakeholder qualification such as this in quadrants will show the project manager some ideas on which stakeholders should receive priority in having expectations and requirements fulfilled. A stakeholder that positions high in the top right quadrant is one the project manager should make sure to invest the appropriate amount of time on in order to meet expectations. Conversely, one situated on the lower left quadrant might not need to receive the same level of attention by the project manager.

Once the stakeholders are qualified by quadrants utilizing the criteria presented above, the project manager should use the following stakeholder management strategy for the stakeholders falling under each quadrant:

Quadrant 1: Keep Satisfied
Quadrant 2: Manage Closely
Quadrant 3: Two-Way Communication
Quadrant 4: Keep Informed

The stakeholders that place in quadrant 3 should be kept informed, but their input should be requested by the project manager on specific topics. The ones placed in quadrant 4 should receive one-way communication on project status. The ones placed in quadrant 1 should receive extra attention by the project manager in a way that their expectations are met. The ones placed in quadrant 2 are the ones who usually have veto power; therefore, they are the ones that should be closely managed.

Table 42 illustrates the suggested format for the stakeholder management strategy.

Stakeholder Management Strategy

Stakeholder Name	Project Role	Influence Interest Product	Supporter (Y/N/EK)	Main Expectation Met (Y/N)	Management Strategy

Table 42. Stakeholder Management Strategy

The stakeholder name and project role are pieces of information that will come straight from the stakeholder register table. The influence interest product is the result of the multiplication of the influence and interest indexes that are also input data from the stakeholder register.

The supporter column basically identifies the stakeholder as a supporter, oppose, or "even keel" (EK) to the execution of the project. It is important to understand your stakeholder audience and know who is for your project, who is against it, and who doesn't care one way or the other. This may save the project manager a lot of aggravation as the project unfolds, and she can set her own internal expectation as to the level of support she can expect from each stakeholder ahead of time.

The main expectation met column shows if the project objectives and high-level requirements are in line, not in line, or partially in line with the specific stakeholder expectation. The management strategy column basically states the high-level plan on how the stakeholder will be managed. This shows the actions that will be taken based on the influence-interest quadrant the stakeholder is placed as well as if the stakeholder is a supporter of the project or not and if the current project scope of work goes in line with the stakeholder's expectation or not.

Examples of management strategies are:
— Invite stakeholder to risk management team meetings
— Send project status report once a week
— Understand why stakeholder is not a project supporter
— Meet regularly with stakeholder to have one-on-one status reports

There isn't really a laundry list on what to do for all possible combinations the table can deliver. The management strategy selected by the project manager for each stakeholder will heavily depend also on the personalities involved; therefore, these strategies should be devised on a case-by-case basis. However, the classification in quadrants as well as the information provided by the stakeholder management strategy table should definitely be used as input to that decision.

Modeling in Support of NPIs

As shown in a previous chapter, the approach to T&M project execution that best lines up with an NPI process is the agile methodology. That chapter also presented the concept of a product backlog, prioritization and ranking of that product backlog, and the concept of sprints.

Since the agile methodology was mentioned, it is worth stressing that what this book actually proposes is a slight modification to agile in its purest form. For that discussion to occur, we need to explain some other concepts of agile and then show why those concepts are not quite appropriate, without modifications, on T&M projects.

Let's go back to the product backlog for a moment for a more detailed account on what it is and how it is created. The product backlog is a prioritized features list, containing short descriptions of all functionality desired in the product, or deliverables. One of the most predominant of agile's tenets is that it is not necessary to start a project with a lengthy account of the project requirements. In fact, the process of creating an agile product backlog is having the project team and product owner write down everything they can think of for the product feature set and prioritization.

The product backlog usually changes and grows as more is learned about the project. Here one can see the first potential problem with applying a purist agile approach to a T&M project, especially if they are being executed by an integrator. As mentioned previously in this book, both clients and integrators like fixed-price T&M projects. Clients like the certainty of project schedule and budget determined up front so these numbers can be plugged into the overall NPI. Integrators like fixed prices when the scope is defined, so they can leverage their own internal tools and efficiency of people who do these types of projects for a living.

The very definition of the creation of an agile product backlog is a showstopper for a fixed-price engagement with an integrator. No integrator would ever fix prices on a project based on a very loose scope of work, defined by an initial product backlog where no rigor was applied in the definition of some level of project requirements that can be translated into a defined scope of work.

Another point that is worth making in this initial discussion is that agile product backlogs usually are created through the utilization of the so-called user story. A user story in agile is very often wrongly confused with use cases. A user story is a very short description of something that a stakeholder will do when they use the system, focused on the value or result they will get from doing it. They are written from the point of view of a person using the system, and written in a language that the stakeholder would use.

A use case is a description of a set of interactions between a system and one or more actors, where an actor can be a person or other systems, as was defined in the last chapter. Use cases are made for requirements to be elicited, whereas user stories are made so the initial, loose product backlog that will kick off the agile process is defined.

With this in mind, it is somewhat easy to see why the pure agile product backlog is not the best way to develop a T&M project that will support an NPI and will be done by an integrator.

However, as was mentioned, user stories are often confused with use cases. The fact of the matter is that use cases can almost be seen as a deeper level of abstraction than user stories. User stories are meant to be superficial and simple. Use cases are meant to be descriptive and all encompassing.

When one adds activity diagrams to further detail the use cases, the initial project requirements can be very closely modeled by just these two diagram types. The whole idea here is the creation of a modified agile product backlog. This modified version of the product backlog will include the use cases and the requirements detailed by the activity diagrams, as opposed to superficial user stories.

There will be a translation process of information coming from the diagrams into a list of tasks, features, and activities, which will be our modified agile product backlog. Once this list is completed, as was mentioned in the previous chapter, the list is then prioritized, ranked, and split into multiple sprints of defined scope.

What this book proposes is a process close to agile from the standpoint that sprints will be executed and potential changes to the original product backlog may come about, due to the fluidity of the NPI process itself. The process is close to agile but not agile per se due to the fact that the product team has much less flexibility to come up with their own changes within the scope of a sprint. There will still be a more formal structure similar to nonagile projects for the implementation of the scope defined by a sprint. However, the nature of this fine granularity breakdown of a large task into small ones that can influence changes on the following ones is very much at the heart of the agile movement.

Once the modified initial product backlog has been created and the sprints identified with their corresponding scopes from the product backlog, the integrator can fix prices for each of them separately as if they were separate projects altogether. This will give the client an initial idea of project budget and schedule that can still be plugged into the NPI master schedule and budget. However, the client needs to understand that this initial project schedule and budget will most likely change due to the fluid nature of the NPI process. This provides the clients with several levers that can be adjusted based on the overall parameters of the NPI.

For instance, NPIs that are not extremely aggressive as far as final product launch can afford to wait longer for the T&M project to start. These types of NPIs will obviously benefit from a more stable device under test when the T&M project is kicked off. This extra level of stability will lead to a more accurate initial product backlog. This more accurate product backlog will lead to an initial project schedule and budget that will most likely have less variation than one where the NPI launch schedule is very aggressive.

In these types of NPIs, the T&M projects absolutely need to be completed at a very near point in time to the completion of the device under test. For those, the initial product backlog will be as unstable as the device under test is at that point in time. This will lead to more variability of project schedule and budget that needs to be taken into account by the NPI management team.

This chapter has focused on the tools element of the people, process, and tools combo. UML has been introduced as well as a few of its diagrams that are very useful in the process of modeling, which is one of the strongest tools that can be used on T&M projects. Project objectives and requirements modeling and system specification modeling were also presented and linked to the types of UML diagrams that are most suited for each activity. Also, a methodology to capture and manage project stakeholders was introduced. This activity is extremely important in addressing the missing stakeholders root cause identified in section 1 of this book. Lastly, it was shown how modeling can be used to support T&M projects that will be part of NPIs.

The next section of this book will focus on the process of people, process, and tools, the TMPM process.

CHAPTER 9:
TMPM: The Process from People, Process, and Tools

The previous two chapters of this section have presented the people and tools elements of the people, process, and tools combo. This chapter will focus on the process element. Since the last chapter presented the tools in detail and how to use them to execute the multiple modeling activities that are part of the tool set of the TMPM framework, this chapter will basically focus on the high-level activities and the corresponding utilization of those tools by each activity. The process for the entire project life cycle will be presented.

It is important to keep in mind that this chapter presents the full set of activities to be performed in a complex T&M project. It is important that the reader understands that this full set can be either executed in its entirety or in part. Common sense is what should be used by the project management body when determining what activities to perform in a given project to avoid unnecessary overhead. Smaller, simpler projects probably won't require the full set of activities described in this chapter.

One last disclaimer before we get going on the process details: this chapter will use the terms project manager and project management body interchangeably. Remember that the chapter on organization structure defined a two-people management team to best manage T&M projects, a systems engineer and a project manager. Therefore, unless specifically expressed in the text, one should assume this two-people management body whenever the term project manager is seen.

Defining the Project Objectives

As has been emphasized in this text, the project objectives are a very important artifact that needs to be generated. It sets the stage for the stakeholders' high-level care-abouts to be addressed by the project requirements. Since bad requirements are one of the most important illnesses of bad projects, and the project requirements are a deeper abstraction layer to the project objectives, having a solid set of project objectives is a fundamental foundation for the unfolding of a successful project.

Referring back to the tools chapter of this book, it presented a methodology for modeling project objectives through the utilization of UML and use case diagrams. As a refresher, use case diagrams define actions that will be performed by actors, as well as define boundaries to the system being modeled. Also, remember that the UML diagrams provide a phenomenal level of flexibility as to what type of information is captured and the several levels of abstraction that it can represent.

One can intuit that if the intention is to focus on the high-level project objectives, use case diagrams can be a very useful first pass at that exercise. Each actor can represent a stakeholder, the use cases can be the stakeholders' care-about actions, and, most importantly, the system boundary can be clearly defined. This is particularly important to make sure the stakeholders understand which of their care-abouts will actually be part of the scope of the upcoming project, and those that will not. This will give them an opportunity to validate that set of assumptions for the system boundary and clarify what is and what is not in scope.

Obviously, each project objective gathering activity will be different from the next. Each one will be particular to the project being modeled. For some simpler, smaller projects, it may be enough to perform the use case diagram analysis presented previously for the full elicitation of the project objectives. However, for more complex, larger T&M projects, the person gathering the project objectives may determine that the use case diagrams don't quite provide all information that needs to be captured. In those cases, the utilization of activity diagrams is very helpful.

One may remember that activity diagrams provide the flow of activities of a given use case. Moreover, that diagram provides a very important piece of information, which is who is responsible for each activity. The diagram has swim lanes with an owner of that lane on top. Every activity that falls within a given swim lane is the responsibility of the owner of that swim lane. It becomes intuitive to the reader at this point that activity diagrams that intend to model the project objectives will map activities under swim lanes that are owned by the project stakeholders. This will provide a very visual representation on how the activities, or in this case, the high-level project objectives, map to each stakeholder as well as the relationship of those objectives among the multiple stakeholders on the project.

This type of diagram makes a great way to represent the project objectives in a meeting with multiple stakeholders with different backgrounds and from different areas on the organization. It presents a nontechnical way of communicating how the project objectives are shaping up, what stakeholders are the ones driving each objective, and how they are intertwined with each other.

The discussion above addresses how to use the tools presented by the previous chapter in eliciting the project objectives. It is important to keep in mind that project objectives have a direct connection with the business motivation driving the need for the project. Having an understanding of the tool set available and how to use them is important. However, probably equally as important is understanding the business environment the organization is currently under and how that drives the need for the project. Remember, project objectives are the statements of value for the project. It is how the organization is expected to benefit, on a business level, by implementing that project. The next sections provide some insight on how to go about understanding the business environment. This information should be used as a guide on how to apply the tools presented in the last chapter.

Understanding the Business Environment

The first step in identifying the business issue to be addressed—either a problem to be solved or an opportunity to be realized—is to understand the environment within which the organization is immersed. The majority of business cases originate from one of the following scenarios:
- New commercial opportunities driven by the organization's target market niche
- Change in overall organization mission, strategy, or objectives
- Reaction to newly launched competitive solutions
- New opportunities derived from the introduction of a new technology
- Overall environmental changes such as statutory, legal, or industry standard modifications
- Continuous improvement and/or operational cost-reduction activities

New Commercial Opportunities Driven by Organization's Target Market Niche

The first high-level business issue from the list above is clearly classified as an opportunity. The only constant factor in the world of business is change; as such, the organizations that are flexible enough to adapt to those changes are the ones that thrive and turn themselves into enterprises.

Apple is a great example of a company that lives by this motto. Apple started as a personal computer manufacturer in 1976. In 1977 it presented its first innovation, a personal computer with color graphics and open architecture, which allowed Apple to officially become a respectable player in that market. In its attempt to grow, Apple decided to jump into the market of business and office computing, and it failed miserably.

Apple learned with this failure and decided to focus on the personal computing market. It realized that the future of this market was in line with the ability of providing a computer with a rich graphical user interface (GUI). It then launched the Macintosh PC, which sold well at the beginning. Macintosh sales started to fade due mainly to its steep price tag and limited range of software titles. Apple realized its strong suite was its graphic appeal; it then launched the software PageMaker, an early desktop publishing package. The marriage between Macintosh's strong graphics capabilities and PageMaker was extremely successful. This combination was probably the biggest factor in the creation of the desktop publishing market.

That market was led by Apple for a few years until Microsoft and its suite of software products transformed the Windows-based PC into a strong desktop publishing market tool. This forced Apple to reinvent itself one more time, as it kept losing market share to Microsoft in a market that was totally owned by Apple in the past.

Apple decided then to invest on digital media and later consumer electronics by slowly transitioning itself out of the PC market and into digital music and video. The result of this switch was the creation of iTunes, iPod, and iPhone. The rest is modern history, and Apple closed 2010 with revenues of $63.5 billion at about a 37 percent gross margin.

Apple's history is a great example of a company that understands its market niches and takes advantage of potential commercial opportunities generated by it. Obviously, the story presented here shows the organization on its highest level of strategic decision making. On a tactical level, these commercial opportunities presented by the market niches and recognized by the organization certainly originated multiple business cases that drove the creation of several project portfolios, programs, and projects.

Change in Overall Organization Mission, Strategy, or Objectives

Let's keep utilizing Apple's history as an example, as it fits this category also. By the early 1990s, Apple was investing in the development of alternative platforms to the Macintosh. The Macintosh platform itself was becoming obsolete as it wasn't built for multitasking and several software routines were programmed directly into the hardware, something that Microsoft had addressed very well by abstracting the PC hardware via software APIs available to the software programmers. In addition to these technical hurdles, Apple was facing fierce competition from OS/2, Unix, and Sun Microsystems.

This made Apple realize that the investment needed for the organization to keep competing in the PC market would need to be astronomical. Also, the market was already occupied by some respectable players, which would make the investment even riskier. This is when the organization almost totally reinvented itself, switching its mission that certainly hovered over the PC market onto another one that would have entertainment and digital media as its main focus.

This change in organization mission also changed the organization strategy, from focusing heavily on selling PC hardware and having a software offering that would be just good enough to push those hardware sales, into having an equally shared focus on both software and hardware. Today, Apple has somewhat of a balanced revenue base coming from both the hardware sales of iPads and other "iPlatforms," and its bubbling App Store, which sells applications for its hardware platforms. The App Store helped to push the Apple ecosystem into something that went viral and now has become some sort of informal standard for digital entertainment.

Much like the topic previously presented around commercial opportunities, this switch in Apple's vision obviously generated several business cases that promoted portfolios, programs, and projects. One last thing worth highlighting is that this change in mission and strategy can be either an opportunity or a problem, depending on the specific business environment in which the organization is immersed.

Reaction to Newly Launched Competitive Solutions

This is definitely a business problem that needs fixing. The snippet of Apple's history showing the change in organization mission as a business issue can also function as an example of the reaction to newly launched competitive solutions. Microsoft's and others' big push to compete in the desktop publishing market forced Apple to end up changing it organization mission as explained above. These launched competitive solutions and most definitely drove Apple to create some business cases to sponsor projects that would help Apple respond to these competitive threats.

The end result was a change in the overall organization mission; however, that is not a decision that an organization the size of Apple takes lightly. One can be sure that several projects and initiatives were sponsored in order to initially try to respond to the threats introduced in the marketplace, followed by other projects trying to establish some alternatives to the ever-decreasing market share, and ultimately, projects that helped determine the new organization mission.

New Opportunities Derived from the Introduction of a New Technology

This definitely qualifies as an opportunity, and Apple can serve again as a good example on this business issue. The new technologies introduced to the cell phone industry are what made possible Apple's success with the iPhone and to a large extent its App Store. Granted, it does take incredible market foresight to practically create a new industry like Apple did; however, none of this would be possible if cell phones were still in the age of voice transmission only.

One can go one level of abstraction higher; the cell phone industry in turn could only take the leap it did, being transformed from being basically a utilitarian tool into an entertainment widget due to technology improvements in the telecommunication industry. The telecommunication industry today provides bandwidths that allow an incredible amount of data to flow through the wireless networks, which in turn allow streams of music, video, voice, e-mail, Internet data, etc. to be integrated into user cell phones.

Obviously, all these new technologies definitely generated an inundation of business cases from multiple organizations that in turn sponsored portfolios, programs, and projects.

Overall Environmental Changes Such as Regulatory, Legal, or Industry Standard Modifications

This segment encompasses any new law passed by governmental entities, changes in regulations by controlling entities, or even changes in industry standards that drive organizations to react to those changes in order to keep compliancy level.

One example of regulatory changes could be any new requirement put in place by the Food and Drug Administration (FDA). The FDA regulates all organizations at play in the pharmaceutical, medical devices, and food industries. Their requirements are a must-have to any organization looking to launch and/or keep a product in the market. A change in FDA requirements definitely triggers business cases to bring existing products and ones under development that are affected by the change into the new compliance standard.

It is somewhat straightforward to understand how new laws passed by government entities can also trigger the creation of business cases. Existing business affected by those new laws might need to fund projects in order to bring their offering into compliance with the new law.

The same can be applicable to industry standards. Industry standards are not necessarily law or regulations sanctioned by official entities; however, they are standards created by the industry community that informally bind the companies that play in that given industry to a set of basic requirements. There are several examples that can be given on this topic. ISO-9001 is a widely used standard controlling the quality level of organizations in many industries. Though organizations are not required to become ISO certified in order to sell their products or services, some of them are pushed by their customer base in order to achieve that level of certification. This big push by the market niche forces those organizations to create business cases that will in turn fund projects to bring the organization into ISO certification.

One can jump to the conclusion that this issue can always be classified as a business problem; however, there might be some cases where this can become an opportunity. Let's look at one quick example illustrating a situation when changes in regulations can become opportunities. Think back to what was probably the most painful day any of us can remember, 9/11. The events that day are forever engraved in our minds. Also, they generated a multitude of new laws and regulations in an attempt to provide a higher level of safety against terrorist acts. One of those changes affected airport security.

The United States approved the utilization of the controversial body scanner at some selected airports. The philosophical implications of this mandate are beyond the scope of this work, but the fact of the matter is that this became a business opportunity for companies that could manufacture the body scanner equipment.

The point to highlight here is that only the environment a given company is immersed in can dictate if these types of changes are business problems or opportunities.

Continuous Improvement and/or Operational Cost-Reduction Activities

One last business issue that is worth mentioning is the continuous improvement and cost-reduction category. The modern business climate is exerting an ever-increasing pressure on all organizations to provide more for less. Today, end users don't expect a cell phone to just allow them to wirelessly communicate with family and friends, but also to stream an unthinkable amount of real-time data, to use it as a high-definition TV, and to surf the web as basic functionality set; and by the way, they expect to pay just a few hundred dollars for their unit. This concept can be expanded beyond cell phones onto a multitude of other examples of the so-called daily life products of our modern society.

This has an obvious impact on how cell phone manufacturers run their businesses. Continuous improvement activities are a reality in today business, as they allow organizations to always look at ways to run their operations more efficiently and at lower costs, so they can survive the market pricing pressures.

Understanding the underlying motivation on a business level for the creation of a T&M project not only will help in the definition of the project objectives more easily and how to best apply the tools presented in the previous chapter, but it will also make sure that the objectives that have been elicited from the group of stakeholders is in line with the top-line business value that is driving the entire project.

Remember, success on a T&M project is not just bringing the project to completion within schedule and budget, but also making sure the project is maximizing the business value to the organization.

As the old saying goes, the first step to fixing a problem is to actually identify the problem through a clear and concise problem statement. This should also be done by the person collecting the project objectives, and will be explained in the next section.

Describing the Business Issue

The previous section presented the concept of identifying organizations' underlying business issues as the first step to creating a sound business case. Once the issue is identified and properly qualified as a problem or opportunity, as well as classified within one of the presented high-level groups, the next step is to describe the business issue in details.

The lists presented below propose basic questions whose answers will guide the person responsible for implementing the business case through the process of creating a detailed description of the issue. This description will, in turn, provide basic information to the business analyst to create the framework of the business case; the impact the problem/opportunity is/will create on the business, and the time frame for putting the solution in place.

The first list is to be applied to business problems and contains the following:

— Reasons why the problem came about
— What business components (human, technology, operational, legal, commercial, etc.) created the problem
— Impact the problem is having on the business (financial, operational, human, legal, etc.)
— Time frame within the problem must be fixed
— Possible consequences if problem is not fixed within identified time frame

Reasons Why the Problem Came About

It is fundamental for the business analyst to perform a very candid analysis and identify the root reasons why the problem surfaced. This information, as painful as it may be for the organization to face its closet skeletons, will allow the business case to highlight the true reasons why the problem started afflicting the company.

The organizational culture must wholeheartedly support this effort, and a "no sacred cows" attitude must start at the highest levels of management and flow all the way down the ranks in order for this exercise to be truly successful. If this is not the case, the business case may be under the risk of barking up the wrong tree, which will ultimately lead to the sponsoring of a project that won't really fix the real problem but instead just masks it just enough for it to keep surfacing over and over.

This information needs to be included as part of the business case. This way, it becomes documented, and whatever projects are spawned from it will always have the end goal in sight. At the end of the day, as it will be seen in later sections, project objectives need to be well established and clarified from the very beginning for the project to have a real chance of success. In fact, the very definition of project success is not only for it to be completed on time and on budget, as most project managers believe; more importantly, the main objectives need to be achieved. A project that is completed on time and on budget but that doesn't achieve the objectives dictated by the business case is nothing more than an expense to the organization and a business failure.

What Business Components Created the Problem?

This topic is, in fact, an extension to the prior one presented. Not only the main root problems need to be identified, but also the root driver for the problem. If the driver of the problem is not properly identified, the business case might sponsor projects addressing the reasons for the problem without addressing what created those reasons. This might not be enough to permanently fix the problem.

The term "business components" presented under this scenario is defined as the multiple components that make up any business: the human, cultural, financial, operational, technological, and commercial components.

One last thing that is important to highlight on this topic is that this exercise is not about finding guilty parties or pointing fingers around the organization. It is important to understand that the responsibility for every business problem is equally shared between the business component that is the source of the problem and the organization as a whole. This is about fixing a problem, not finding sacrificial lambs.

Impact The Problem Is Having on the Business

This is another very important data point that needs to be captured by the business case. The end goal of this piece of information is to allow the organization's high management to have a snapshot of the financial implications created by the business problem at hand.

It is important to always keep the last statement in focus. The simplest of the financial equations, where gross margin is defined as revenue minus cost of obtaining the revenue, is what drives the majority of business decisions, and funding of projects is not different. A business case that will sponsor a project needs to show why the project is needed in the first place. In the case of a project to fix a business problem, the justification for creating the project needs to show the cost of not doing the project, that is, the financial implication of the problem, or, in yet other terms, how much this specific problem is costing the organization.

When this number is quantified, one can start showing the return of investment (ROI) of the project to be funded to fix the problem. The project ROI will be basically the time it will take the organization to recover the money used to fund the project, or the project budget, once the business problem is addressed by the project. Later sections will cover ROI in more detail; however, a quick introduction on the topic was appropriate as part of this section in order to better explain the reason for this data point as part of the business case.

Time Frame Within the Problem Must Be Fixed

This is really what the statement says. Some business problems are so severe that they can potentially force the organization out of business altogether. This would be the extreme case. More intermediate cases would be business problems that would prevent the organization from reaching specific set goals. Regardless of the final impact of business problems, all of them can be seen as time bombs to the business, and the exact time of the bomb clock needs to be determined.

Much like the case of a real time bomb where the bomb squad needs to know how much time they have before the bomb goes off, a business problem time frame until catastrophe needs to be quantified.

The underlying reason is somewhat straightforward. Once the organization's upper management understands the time frame of a specific business problem to be fixed, the business case that will sponsor projects to fix the given problem will include the information so the project, once created, will have a clear definition of the schedule objective. A project that is completed after the time frame of a business problem has expired doesn't achieve the business objective that drove its creation and, therefore, fails.

Possible Consequences If Problem Is Not Fixed Within Identified Time Frame

The previous section mentioned that the business time frame needs to be determined and that some problems can potentially drive an organization out of business. This being the extreme case, the consequence of a problem of such nature is obviously the death of the business. A project that is created to address a business problem of this nature obviously need to have higher priority than any other parallel project as well as receive all of the organization's focus and resources. In the middle of the spectrum, though, things are not as black and white.

Business problems are a constant in any organization, and though organizations may not be facing imminent extermination, some problems are more serious than others. It is paramount that the organization understand the consequences to the business in case problems are not fixed within the determined time frame. This information will be used by the executive team in order to prioritize projects within the organization. Common sense says that problems with more serious consequences need to be prioritized over problems of lesser consequences. Though this is obvious when put under this framework, in the real world of business, determining what problems carry more drastic consequences than others is not as easy to do as one may think; therefore, focus and attention need to be given to such a fundamental topic.

A business case that clearly specifies what happens if projects are not funded and when they absolutely need to be funded will have a much greater chance of being selected to become an active business case in the organization than others that don't present the same level of detail. Executive management of any organization thinks dollars and cents, and speaking the language of management is very important for any project manager. A project manager doesn't need to become a financial wizard, but knowing the basics of business accounting and financials certainly helps someone managing projects. If this is not a topic you have some familiarity with and you now see yourself in a project manager capacity, it would be strongly advisable to obtain some fluency in the topic. It will help not only in your communications and status reports to your management team, but will also help you on a more tactical level when you manage your projects.

This section so far has detailed the questions one needs to ask and obtain answers to when describing a business problem. Let's now switch gears and address the questions that are to be asked and answered when business opportunities are to be made into business cases.

The list below is to be applied to a business opportunity:
- Describe the business opportunity in details
- Describe why and how the opportunity flourished
- Describe the opportunity window, that is, the time frame the opportunity will be available
- Define the impact that realizing the opportunity will have on the business
- Possible consequences to the business if opportunity is not realized

Describe the Business Opportunity in Detail

This section of the business case is where the business analyst can present the opportunity to higher management. The more information included in this section the better, as it will show the decision makers that the appropriate research and due diligence was applied by the business analyst before putting the opportunity in front of management for a decision.

There isn't really a defined laundry list of things that need to be covered as part of this section, and the type of information presented will very much depend on the environment the business is immersed in. The main thing to keep in mind when creating this section of the business case is that this section needs to include all needed information in order to present the opportunity to the organization stakeholders.

Describe Why and How the Opportunity Flourished

This can be seen as the background information showing the known reasons why this opportunity is now available to the organization. This is part of the appropriate due diligence required when researching a given business opportunity and will, along with the information provided by the previous section, be appreciated by senior management when presented with the opportunity.

Since market analysis is as much of an art as it is an analytical science, and because very seldom the entire data set needed for determination of the opportunity is available, assumptions need to be used as part of the business case. The documentation of these assumptions is important so senior management can have a chance to validate the assumptions used for the presentation of the opportunity, prior to making the decision of approving or denying the business case.

Describe the Opportunity Window

Much like in the case of a business problem where it is paramount for the business analyst to present the time frame within which the business problem needs to be addressed, the business opportunity has a counterpart, which is called the opportunity window.

No opportunity remains open forever. In today's competitive business climate, one of the main drives of all organizations is the so-called concept of time to market. It is a known business fact that the highest margins are commanded by product innovators and the first one to get to any market; therefore, it is easy to understand that the less time to market in any product launch, the higher are the chances for the organization to enjoy that position for the longest time possible.

Any new product launch, as an example, has an opportunity window, which can be the time until a competitor product is launched, the time frame the market environment will remain favorable for the new product, the time a specific gross margin can be commanded by a new product, etc.

Serious market analysis needs to be executed in order for the opportunity window to be identified. This information, much like in the case of the time frame for a business problem to be solved, needs to be documented as part of the business case. This information will help senior management prioritize the multiple business cases and potentially approve the sponsoring of projects. It will also be used by the project manager to determine the schedule objective for the project.

Define the Impact That Realizing the Opportunity Will Have on the Business

The information presented in this section, in very high-level terms, should ultimately show the revenue schedule that the new opportunity will bring to the organization. In lower-level terms, information showing how this revenue will be brought in by the opportunity should be presented. This information should be as data driven as possible, and the data based on verifiable market facts. This will increase the confidence level by senior management when analyzing the business case.

The size of the opportunity—that is, the amount of revenue that it will bring to the organization—is one of the inputs in the calculation of the business case ROI, the other one being the business case budget. The notion of ROI was presented in previous sections and is applicable in this case also. The size of the opportunity might be large; however, if the cost of realizing the opportunity is larger, the business case is a loser and shouldn't be approved. This calculation will be an input to the decision of approving the business case or not.

Beyond the yes/no decision by senior management about the business case, if the business case is indeed approved, another decision can be made by using the information about the impact of the opportunity to the business, which is the project budget. This information will be rolled into the project and used by the project manager as one of the project objectives to be achieved by the project.

Possible Consequences to the Business If Opportunity Is Not Realized

The analysis done during the previous section will provide the size of the opportunity if realized. However, another very important data point to be used during the decision-making process of approving a business case is the consequence—the cost—if the opportunity is not realized.

As an example, sometimes an opportunity might not be so large if analyzed in isolation to justify a business case approval; however, if the cost of not realizing it is taken into account and added to the formula, the overall financial picture is totally different and the business case will be approved. One simple example that comes to mind is of an organization that is already the market leader of a segment and is analyzing the possibility of launching its brand in a not so large international market in relation to the overall brand revenue numbers. The added revenue to the overall brand wouldn't be large enough to justify the investment; however, the organization's main competitor is a known player of that international market and has been aggressively launching new products on that market. If left unattended, the competing organization can grow its overall strength by driving that international market.

The scenario above shows that there might be some unwanted consequences if the opportunity is not pursued, and that these consequences can cost much more to the organization than the potential cost of the projects to realize the opportunity.

As previously stated in this text, this account assumes a very large and complex T&M project that is rooted in a business issue that needs to be solved, which is motivating the funding of the project. Obviously, the implementation of what was presented above would be overkill for smaller projects. The person performing the gathering of the project objectives should use best judgment on the level of detail that will be invested in the analysis. What this book does propose, however, is that project objectives should always be captured, regardless of how simple the project being initiated seems to be.

Once the exercise above has been initiated and is under way, another tool, presented in the previous chapter, that should be applied in this process is the creation of a stakeholder registry along with the creation of a stakeholder management plan.

Hopefully, the high-level modeling exercise facilitated the identification of a large group of stakeholders that are somewhat touched by the project. Also, the modeling activity helps in identifying information other than just who the stakeholders are that are important for the stakeholder management plan to be put in place, such as level of influence and level of interest. Usually, during the process of involving the stakeholders in these modeling activities, the stakeholder names that are always being thrown around by other stakeholders become clear. Those are probably the high influencers. Also, the person leading the modeling activities can tell fairly quickly what stakeholders are interested in the success of the process and which ones don't really care that much. This will help in qualifying the stakeholder's level of interest.

The next level down the project life cycle should be the creation of a project charter, which is the subject of the next section.

The Project Charter

Now that the project objectives have been identified, it is time to start working on the project itself.

The project charter is a document that formally authorizes a project or phase and documents the high-level initial requirements that satisfy the stakeholders' needs and expectations. This document, ideally, should be formally authorized by a representative of the organization management team. The approver of the project charter will be known hereafter as the project sponsor.

The project charter will summarize the following:
- The high-level project objectives
- The high-level risks
- The utilized project assumptions
- The project acceptance criteria

Project Objectives

The first step toward creating a sound project charter is to detail the high-level project objectives a little further. Everything done in the project from this point onward needs to be focused on the project objectives, as they are the sole reason for why this project came into existence in the first place.

Just as the business case presented the highest level of the objectives to be met, the project high-level objectives will present the next tactical level of objectives the project needs to achieve, and that will include the business objectives.

Project Vision Statement

To ensure that all project team members are working toward a common goal, the project needs to have a vision statement. This vision statement will be short so that all project team members can easily memorize it. At the same time it will capture the essence of the high-level business objectives as stated in the implementation plan section of the business plan. This may sound silly at first; however, do spend the needed time on creating the proper project vision statement as everything else from then on will be generated from these few words.

Project Objectives

Based on the vision statement, and assuming the vision is indeed encompassing all the high-level business objectives, list the identified project objectives in a SMART format. The acronym SMART stands for specific, measurable, achievable, realistic, and time-bound. Remember that at the end of the project, there should be a way to verify that these objectives were met. In order for them to be verifiable, the project objectives need to be SMART.

One tip for writing these objectives is to start thinking about how you will be able to actually demonstrate that the project has achieved them. Vague terms such as "improve" or "enhance" need to be followed by analytical numbers, such as "to improve X by Y percent" or to "enhance X by a factor of five." This will allow an unambiguous methodology to verify project completion and a way to better measure project performance along the way to verify if the project is indeed converging to an end or if it is out of control.

Project Deliverables

Much like the case of the business case where the expected business deliverables were listed as part of the final summary table, the project charter, as it will be seen in the next chapter, should also capture the deliverables, but on the project level. As mentioned, the project charter is the highest-level project document, though it is a child document generated from the business case.

Table 43 can be used as an aid to creating the project level deliverables.

Project Deliverables

Deliverable Name	Components	Description

Table 43. Project Deliverables Table

The components column describes the major components that constitute the corresponding deliverable. For example, all the documents, new software, prototype, and any other artifact that is created to constitute the deliverable is a component. The description column will include a detailed description of each deliverable so anyone who reads the project charter can understand what will be generated. Always keep in mind that people from a multitude of backgrounds can read the project charter. As such, the verbiage needs to make sure there is no special jargon that would make it difficult for people with a different background to understand what the document means.

High-Level Risks

Risk, even though it is a word most project managers don't like to say, can make or break every project, and was presented as an underlying issue from the poor planning root cause. A project that has an appropriate risk identification and mitigation planning activity is usually one that is low stress and more easily managed. The ones that don't go through the same level of risk analysis are the ones that keep the project team up working to 10:00 p.m. and on weekends only to still run beyond schedule and over budget.

The main suggestion here is to not skimp on this step. Do not face this as just a checkbox for something you have to do in order to get project charter approval. The best project managers spend about half of their time doing risk identification and analysis, and that time is always time well spent.

Table 44 shows an example that can be used to support risk analysis activities.

High-Level Project Risks

Risk Description	Risk Probability (0–100%)	Risk Impact (1–5)	Risk Cost ($)	Possible Risk Mitigation

Table 44. High-Level Project Risks Table

As a refresher, risk cost is the amount of money that would a realized risk would cost the project. Enter a good description for each identified risk, its corresponding probability of happening, its impact (one being the least impact and five being the highest), and possible risk mitigation actions that can be taken to mitigate each one of the identified risks.

It is important that sound risk mitigation actions be put in place for the risks that rank higher on the risk probability impact product. This product is the simple multiplication of the risk probability by the risk impact. The higher the number, the more likely the risk will seriously affect the project should it be realized. Therefore, the more up-front planning that can be done for the risk to be mitigated, the better off the project will be.

This exercise will also be useful because it can flag the land mines that the project needs to steer clear of. Knowing the areas of risk of allows the project team to potentially design solutions that stay away from those areas as much as possible, even avoiding them altogether. Not knowing where those mines are will create a component of luck that is not desirable when managing a project. As Murphy would say, "Anything that can go wrong, will go wrong, and at the worse possible moment."

The idea is that even though risk was captured as part of the project charter at a very early stage of the project life cycle, the project manager should never stop identifying and analyzing risks for the entire duration of the project.

Project Assumptions

Now that the project objectives and the initial risks have been identified, it is very important for the assumptions that were used in creating those sections to be documented as part of the project charter also.

Since the project charter is a document that will be signed by the project sponsor, making the assumptions part of the document will force another level of review. The cost of fixing an issue grows exponentially between the multiple phases of a project life cycle, so the sooner a problem is spotted, the cheaper it is to fix it. A bad assumption, if not caught at an early stage, can potentially escalate into several bad project artifacts and even dictate an entire wrong direction for the project to take.

This suggests that any bad assumption that is not caught at this stage can turn into a very costly mistake at later phases of the project life cycle. It is important, therefore, that absolutely all assumptions be captured and transcribed into the project charter, as silly as they might sound.

Simple commonsense assumptions—for instance, the organization has as part of its staff a combination of all needed skills to execute the project—need to be properly verified. An assumption such as this can fall apart if the resources are being utilized in other projects and won't have the needed availability to work on the project for which the project charter is being created. Something simple like this might add a considerable delay if the organization has to hire new people to complete the staffing of the project after it has started.

Acceptance Criteria

Another very important section of the project charter is the definition of the actual project completion criteria. Remember, the project objectives were listed as part of the project charter, and they were made into SMART objectives. It was mentioned at the time that the main reason for them to be SMART was so they could be properly verified.

The acceptance criteria, if met by the project deliverables, basically identify the project as successful and complete. It is important to keep this list of criteria in mind also for the planning activities as the level of quality required in order to complete a project can add a significant schedule and budget overhead to the overall project.

Once the quality standards are defined and agreed upon in advance, the planning activities will have to take those standards into consideration in the project plan. Imagine, for instance, one project where its quality standards must be compliant with strict FDA (Food and Drug Administration) regulations and a second project where no regulatory agency will be involved. The project plan for the first project will have to include a much more comprehensive verification and validation plan than the one for the second project, affecting its overall schedule, cost, resources, risk, and other parameters that are considered during the project planning activity. Now imagine if the acceptance criteria for the first project fail to capture that the project needs to withstand a FDA standard quality audit. This is a really gross thing to miss, but you get the idea.

Should this distinction not be known prior to the project planning activity being kicked off, the project plan will be created based on either an assumed quality standard by the project manager and/or the project team, or one that is probably ambiguous. This will most likely lead to an eventual rebaselining of the project at some point during its execution, leading to project failure.

Table 45 can aid in the creation of sound acceptance criteria for the project.

Project Acceptance Criteria

Project Deliverable	Success Criteria	Verification Method
Deliverable #1		
Deliverable #2		
Deliverable #3		
Deliverable #4		

Table 45. Project Acceptance Criteria Table

The success criteria describe basically the analytical definition of what an acceptable deliverable would need to perform. Feel free to include as many lines per deliverable as needed. It is important to keep a good level of traceability to all the deliverable components for deliverables that are producing multiple outputs. Each deliverable must include independent and analytical success criteria.

The last column to the right indicates the method by which each deliverable will be verified to the success criteria. Examples of verification methods are verification by test, inspection, comparison, association, and independent audit.

Requirements Gathering

As presented in the tool chapter of section 2, requirements modeling is a great way to address the poor communication with stakeholders and missing stakeholder issues. Also, it provides a systemic mechanism to elicit requirements that reduces errors of omission and the syndrome known as "users don't know what they want."

The main idea is the utilization of use case and activity diagrams, as presented in the aforementioned chapter, but now with focus on the project requirements as opposed to focus on the project objectives as the previous section proposed. The user can refer to the section named Modeling Project Objectives and Requirements for a memory refresh on the approach.

What this book proposes is not the abandonment of the typical project requirements document. On the contrary, as will be seen a little later in this section, a document is needed in order for a traceability matrix to be created. What the framework proposes is that the requirements-elicitation activity shouldn't use the typical text-based project requirements as a tool to gather requirements. As seen in the chapter about tools, the method of collecting requirements via text has several problems, some of them in line with the problems that were identified in this book as drivers of T&M project failure.

What the framework proposes is for the requirements-modeling activity, as presented in the chapter about tools, to be used as the main methodology to gather requirements. Once these requirements are gathered to the extent the project management body deems necessary, a requirements document will be created, as the translation of what the diagrams from the models have captured.

Traceability Matrix

Traceability matrix is an artifact that is extremely important on larger projects. In a simplistic way of explaining it, the traceability matrix is a table where higher levels of abstraction artifacts are traced to lower-level artifacts. For example, it was mentioned in this text that the project objectives are a higher level of abstraction to project requirements, or in other words, the project requirements will come to fruition in order to address the high-level project objectives.

Expanding this line of thinking through other phases of the project life cycle, the design would be the next natural deeper layer of abstraction. The system design needs to address all project requirements, no more, no less. If a design actually implements more than the project requirements call for, the project team is doing more than it should, that is, feature creep. As such, we ought to have a mechanism to make sure that all project requirements are being addressed by the proposed design and that all design items are addressing only existing approved requirements.

Going down one more layer, the test plan created to make sure the implementation is correct needs to address all design elements, in order to provide full test coverage. If a design item is missed in regard to verification, usually what ends up happening is that problems will be uncovered at deployment time, or, even worse, after the system has been released for production. Both are huge issues for the project since, as mentioned previously, the later in the project life cycle issues are uncovered the more expensive it is for them to be fixed.

One way to express such relationship between artifacts of different abstraction levels is the traceability matrix. Ideally, all phases of the project life cycle and their main artifacts should be traced all the way up, back to the highest level of abstraction in the project, the project objectives. Figure 35 illustrates the concept.

A traceability matrix that provides full coverage of all artifacts up to their predecessor in level of abstraction is communicating the message that all project objectives identified have been taken into account when the project requirements were derived, that every project requirement is being addressed by the proposed system design, and, ultimately, that all design elements are being properly tested and verified prior to final system delivery.

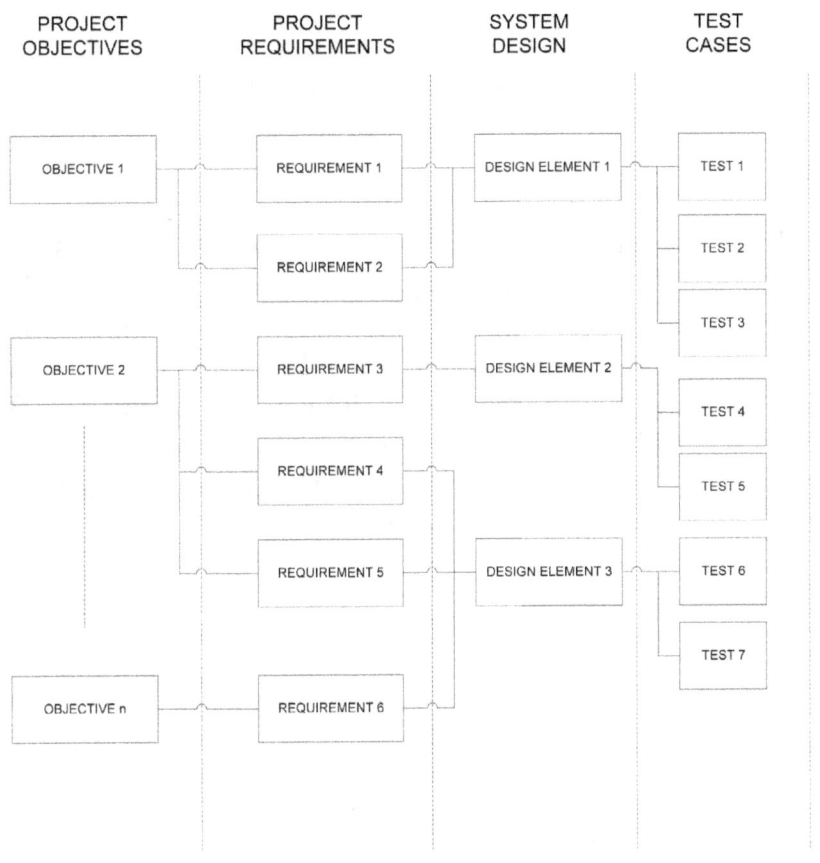

Figure 35. Traceability Matrix Concept

Project Planning

Now that the project objectives have been collected and the project charter created, it is time for the project plan activities to start. Keep in mind that bad requirements and poor planning are the top two problems on complex T&M projects. Therefore, the content presented in this section also carries a lot of weight in whether the project will be a success or failure, as it addresses the planning activities.

The first aspect of the project plan discussed in this section is the work breakdown structure, or WBS.

Work Breakdown Structure

Many project managers make a laundry list of tasks that need to be done within the scope of the project and call that list the work breakdown structure (WBS). This is a huge mistake. The task list, or activity list as it is known in project management circles, is a process that is done after the overall WBS has been defined and is used as an input to the resource planning process. In fact, the WBS is an input to several planning activities as illustrated by figure 36.

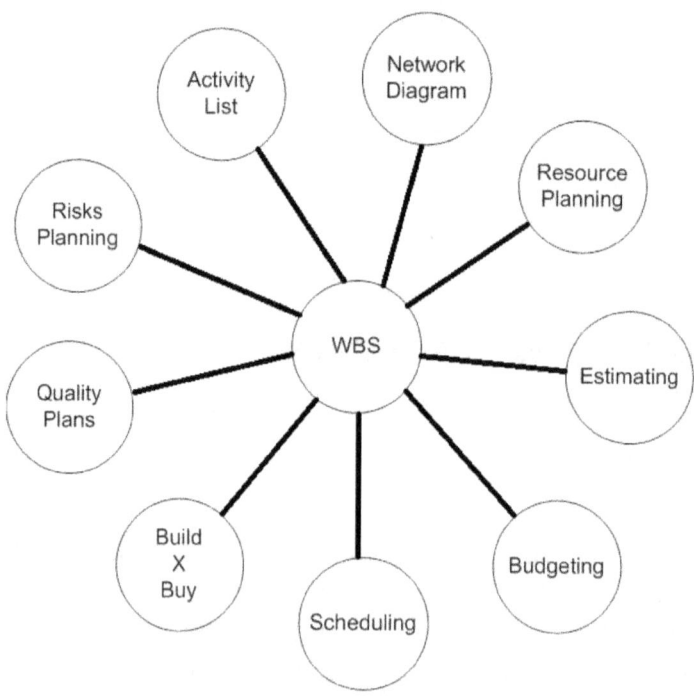

Figure 36. WBS as Input to Other Planning Processes

The WBS is thus a diagram, much like an organizational chart diagram, showing all the work packages that are included as part of the project as well as their hierarchy in relation to each other. Figure 37 illustrates the format of the WBS.

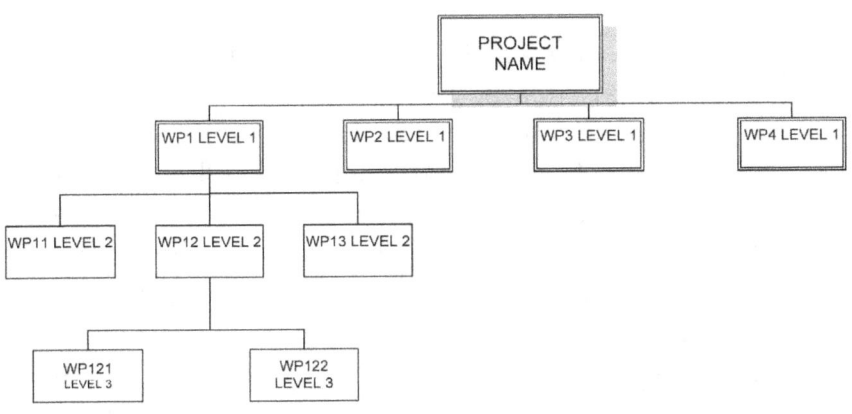

Figure 37. WBS Hierarchy

The hierarchy starts with the project name on top and the higher-level work packages (WPs) taking level one of the hierarchy. The level-one work packages are usually the phases of the project life cycle and come straight from the specific framework utilized by the organization. Level two and beyond are the levels that present the WPs needed for each one of the project life cycle phases to be performed and properly executed. The diagram above shows this exercise done on only one of the highest-level WPs, but on a real WBS, all high-level WPs or project phases obviously need to be dissected and broken down into smaller WPs. There aren't a fixed number of levels that specific project phases need to have. A great rule of thumb is that each of the lowest-level WPs should not include work that takes over eighty hours to complete. This is basically a top-down approach to detailing the project.

One other important practice to highlight is the inclusion and identification of the deliverables as a WP element. This will allow the project manager to build the schedule in a later planning process where those deliverables can be easily made into milestones and/or properly identified in the schedule.

The WBS presented in this format is better than the task list approach due to the following:

- The construction of the WBS on this format allows the project team to have better visibility of the WPs, reducing errors of omission created by the task list approach
- A list is usually created by a single person, whereby the WBS is a brainstorming facilitation for the project team to contribute to
- The team exercise will allow the project team members to better understand the overall project and potentially catch risks that weren't identified before
- Task lists don't allow WPs to be properly broken down into the eighty-hour rule of thumb

One last thing worth mentioning is that it is good practice to number each of the WPs for ease of location later. This might not seem like such a big deal now, but imagine a large project with thousands and thousands of WPs. Having a numbering scheme that notes the location and the interdependency of the WPs will be of immense help later in the project. A system like the numbering scheme of sections and subsections in a Word document is a suggestion.

If it hasn't been clear thus far, it will be made clear now: the creation of the WBS is supposed to be a project team activity, and *not* a project manager activity. Many project managers fall into the trap of trying to do the WBS themselves and end up creating a bad foundation for the project planning as important things are missed by not including the domain experts in the process. Subject expertise is extremely important, as is the power of brainstorming. Multiple minds are always more powerful than a single mind in any circumstance.

Again, it is important to understand the difference between a work package and an activity. A WP may contain several tasks or activities. The activities are the ones that eventually make it into project scheduling software packages; however, the WBS is the source of them all. Also, WPs do *not* show any potential dependencies of either resources or activities. They are merely work and/or deliverables that need to be produced as part of the project scope. They can be seen as small projects in themselves.

WBS Dictionary

Another very important concept in project management that is not very well utilized by novice project managers is the so-called WBS dictionary. In figure 37, the WBS hierarchy, imagine a work package name such as "Build User Interface." This name is not very descriptive of the scope of the WP. It only hints in very high-level terms what will be done as part of its activities. However, the WP name is not easily assignable to a team member without a more detailed description of what the deliverables and/or scope of this activity really are. This is exactly the role of the WBS dictionary.

The WBS dictionary provides a scope definition for each of the WPs so that decision is not left to the imagination of the project members. If this is not done, scope creep will most likely happen, and the project manager will have to take reactive actions to them as opposed to preventing them from happening altogether by having a clear scope definition to be handed off to the project staff responsible for producing that piece of work. Remember, good project management is all about preventing things from happening, other than reacting to fix bad things that actually happen.

Table 46 shows an example of an entry in the WBS dictionary for a single WP. The entire WBS dictionary will contain as many items as there are WPs.

WBS DICTIONARY: WP ENTRY	
WP General Information	
WP Number:	
Responsible Individual:	
WP Description:	
Acceptance Criteria:	
Deliverables:	
Assumptions:	
WP Planning Information	
Resources Assigned:	
Estimated Duration:	
Estimated Cost:	
Due Date:	
Before This WP:	
After This WP:	
Revision History	
Current Revision:	
Date Modified:	
Modified by:	

Table 46. Example of WBS Dictionary Entry

Each of the fields is explained below.

WP General Information: The WP general information section includes the information that is to be filled in at the creation of the work package WBS entry.

WP Number: The identification number of the corresponding WP in the WBS hierarchy.

Responsible Individual: Studies show that every work output expected by project team members, even if they are to be executed by a team of people, needs to have a clearly identified single owner. The responsible individual is the one who is assigned to make sure the work needed by this WP gets done to meet its specified requirements and within the needed time frame.

WP Description: This is the field where details specifying what this WP is and what it does are entered. It is fundamental for the project manager to enter detailed enough information so scope creep is prevented, by having the project teams involved in this work understanding exactly what the expectation for this WP is.

Acceptance Criteria: Much like the project as a whole that has overarching acceptance criteria determining what the project needs to accomplish and at what quality, each WP should have similar information in a microlevel. This information will guide the assigned project team members to understand when the WP can be called done.

Deliverables: As the name suggest, this will be the field where all the expected deliverables for this WP will be listed.

Assumptions: This field will capture all the assumptions that were utilized in the creation of this WP. Much like the case of the overall project assumptions, it is also very important that the assumptions used on the level of the WP are formally captured. This will allow the project team members assigned to the corresponding WP to speak up early in case the utilized assumptions don't jive for whatever reason. This, in turn gives the project manager the opportunity to take preventative actions as opposed to reacting to the snowball effect of having to integrate a WP that was built under a bad set of assumptions.

WP Planning Information: This information is to be filled in once the planning activities are more detailed and decisions are made on what resources will be applied to the WP, what is the deadline for the WP completion, estimated cost and duration for the work, and WP dependencies given by the network diagram.

Resources Assigned: This information will come from the completion of the resource allocation processes and is important to be included as part of the WBS dictionary so there is no confusion on who will be assigned to what WPs.

Estimated Duration: This information is given by the resources assigned to the WP and validated by other domain experts and should be included as part of the WBS dictionary in order to make it easier for the project team members assigned to the work to manage their own progress as well as for the project manager to monitor work performance.

Estimated Cost: This information will come from other planning processes to happen farther down the road. It will let the project manager monitor the project budget performance on a WP level.

Due Date: This information will come from the project schedule once it is created and will help the project team members assigned to the WP monitor their own progress as well as for the project manager to easily identify WPs that are under risk of pushing the schedule out.

Before/After This WP: This information defines the WPs that have some sort of dependency with the current WP, either being predecessor or successor WPs. This information is important as it provides easily accessible information on the project team members for the dependent WPs in case they need to be contacted with questions related to the integration of the current WP with the dependent ones.

Revision History: This information captures the date of the last modification done to the current WP as well as the current revision and who made the change. This allows for traceability of the work throughout the entire project execution.

Activity List

As mentioned in the previous section, the WBS is not a list of tasks to be executed by the project team. The activity list is that set of tasks. This list is obtained through the decomposition of each WP captured in the WBS onto their corresponding lower-level activities.

The smaller the activities the greater the accuracy on estimating and scheduling; however the trade-off is that the higher the granularity level the harder and more time consuming the task becomes. Also, the lack of project knowledge at this early stage in the project life cycle might prevent the project team from parsing the activities in the needed level of details for this to reach its maximum accuracy potential. In these cases, the project manager usually decides to plan on the higher level and come back to the smaller levels of planning details later in the project life cycle. This is called rolling wave planning. The project manager, though, should *never postpone planning* and use the lack of knowledge in the beginning of the project as an excuse to do so.

When completed, each WP will contain a set of activities that when combined will make up for the entire activity list of the project.

Once the activity list is completed, the next step is sequence them into how the work will be performed. This activity is usually done by the meaning of the so-called network diagram. In high-level terms, the network diagram will look something like that shown in figure 38.

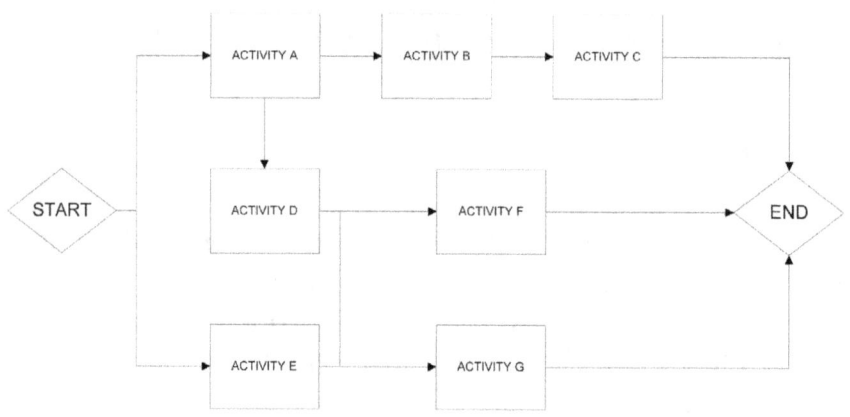

Figure 38. Example of Network Diagram

Each node of the diagram above is used to represent a single activity from the activity list, and the arrows are used to show the dependencies between the activities. In the diagram, it is clearly shown that the project starts with activities A and E that can be performed in parallel, which means they don't carry any dependency. Activities B and D can start only when activity A is completed, activity F can start only when activities D and E are completed, and so on.

In addition to the dependencies between the activities, more information can be entered in the diagram, such as leads and lags. A lead may be added to the start of an activity before the predecessor activity is completed. For instance, in the diagram above, if a lead of three days were included at the start of activity D, this activity could have been started three days prior to activity A's completion. A lag is waiting time that is inserted between activities. For instance, if a lag of four days were included at the start of activity C, this activity would need to wait four days after activity B was completed before starting.

The network diagram is the project manager's and project team's best tool for the processes of estimating resource requirements, time, and cost and developing the project schedule and budget. The network diagram is an exercise to be carried out by the entire project team and not just the project manager, and it allows the team to have a good visual of how the activities should come together to implement the WPs and ultimately the project as a whole. This visual exercise brings the following benefits:

- Shows interdependencies of all activities
- Shows the workflow, which will help properly sequence the activities
- Highlights opportunities for schedule compression during the planning processes as well as the entire project life cycle

Once the network diagram is complete, the next natural step would be for the project management body to determine the final resource requirements for the project and determine how many of each resource type will be required in order for the project to be completed within the schedule stated by its charter.

Project Resources

Now that the activities are listed and sequenced in a network diagram format so all dependencies are captured, the next step is for the type and quantities of the resources to be determined. It is important to highlight that resources include project members but also show equipment and material. Resources can be defined as anything that is needed for the activities to be carried out.

Resources need to be properly planned and coordinated in order to avoid resource shortages when the activities are to be carried out. For instance, a given activity might be the only one that requires a specific type of equipment in order for it to be executed. It is important the project allocates that equipment at around the time it will be needed so other competing projects and activities don't do the same in the same time frame. If that is not done, the activity might be delayed due to the equipment being in use by another project, potentially creating delays in the entire project schedule.

Based on the network diagram, a list showing all needed resources and the time frame needed by the project should be compiled. Expert judgment from domain experts should be used in compiling such list. Table 47 shows a suggestion on how to properly identify the project resources.

PROJECT RESOURCES					
Resource Name	Resource Basic Skills	When Needed (From, To)	Total Hours Needed	Resource Available (Y/N)	Person Name

Table 47. Project Resources Table

The resource name will be the high-level identifier, such as quality engineering, software developer, electrician, etc. The resource basic skills column is for some level of detail on a specific skill set that is required for the activity to be performed. For example, for a software developer, basic skills might be "five years of experience with .NET platform and SQL Server."

The when needed column shows the time frame within which the resource will be involved in the project. The total hours needed column shows estimated hours that the resource will be involved in the project. This information will allow the project manager to determine if the resource needs to be working on this project full time within the stated time frame, or if it is acceptable that the resource be utilized on a different project simultaneously. This information will help the project manager on his hunt for internal resources, which is displayed as a yes/no answer to whether there is a resource available within the organization or. The person name column displays the name of the person who was identified to function as a specific resource.

With this information in hand, the project manager will need to take a second pass at the network diagram, paying special attention to the cases where activities are set to be executed in parallel and require resources with same basic skills of resources that are marked as not available on the table above.

The project manager should attempt to change the network diagram to account for that lack of resources, trying to accommodate the project to the available resource pool. The obvious constraint is the project schedule information set at the project charter. If the project manager can creatively shuffle the activities in order to need only the resources that are part of the available resource pool, the next step is to actually secure those resources by speaking with their corresponding functional managers. If this can't be accomplished, the resources will need to be acquired from the outside, either via contractors or supplier organizations. This will thus serve as an input to the procurement planning processes.

Ultimately, in the later case, it always helps to revisit the project assumptions in the project charter and look for assumptions made around the project resources. If there are assumptions that only internal resources will be used, for example, the lack of needed resources is indicating that assumption is false, and therefore the project charter will need to be reviewed. For instance, this assumption might lead to a project budget set in the project charter that doesn't take into account the cost of external resources. In this case, the project sponsor needs to be involved, as either the project budget needs to be modified to include the utilization of external resources, or the project needs to be killed altogether if the business case doesn't support the new project budget.

This simple use case was presented to highlight how the project documents are interrelated and how they are living entities throughout the life of the project. A decision that was made in an early project management process is not set in stone and needs to be constantly validated in the light of new information or new development. In the case above, it is much better for the organization to kill the project at this early stage should the new project budget not be supported by the business case than at a later stage when much more money has been sunk into the losing project. On project management, knowledge is definitely power, but early knowledge rules.

Estimation

In a properly managed project, there is a WBS with a complementary WBS dictionary that the project team helped create. Also, there is an activity list and a network diagram that came about from the WBS and WBS dictionary and was also codeveloped by the members of the project team. In this circumstance, the corresponding domain experts should not have a problem estimating the time that each activity should take for completion. It is very important to stress the fact that time estimates for the activities should *not* be done by the project manager, but by the domain expert that will be the one performing the activity.

Having the activity owner estimating the time the activity will take to be implemented carries the following advantages:
- It achieves immediate buy-in by the activity owner
- It avoids further excuses by the activity owner stating the activity wasn't completed on time since the time estimated was done by somebody else
- It brings a sense of shared ownership in the overall project schedule as the activity time estimates were done by the project team

Conversely, on a poorly managed project, either the artifacts mentioned above are not created or they were not the result of a team effort performed by the members of the project team. When asked to estimate an activity under this scenario, it is not uncommon to hear the activity owner stating, "I don't know the scope of what I need to estimate so I will take a wild guess and pad it."

Padding, a common practice, is one that can lead to disaster. Imagine if all activities are estimated with padding. The project will end up with a bloated schedule and cost as project baseline. Some project managers say they should underpromise and overdeliver, and that if they can pull off a bloated baseline and deliver under budget and ahead of schedule, they will look like heroes. This is extremely bad project management. Remember that every project is supported by a business case, and on the business level, decisions should be made based on as accurate information as possible. An organization that uses padding as part of its project management culture might be passing up opportunities to start projects that would beneficial to the business due to inflated schedules/budgets being communicated by project managers.

There is a concept in project management that will be presented in later sections called project variance. In high-level terms, project variance is the difference between the current state of the project and its baselines. Ideally, the variance should be as close to zero as possible at the end of the project. A large variance means that either the project was poorly managed and ended up over budget and behind schedule, or it was padded on its estimates and finished way under budget and under schedule.

The best project managers, the ones that should be perceived as true heroes by the organization, are the ones who consistently bring their projects successfully to completion with small variances. With this in mind, padding should be eliminated from any estimating practice. For that to happen, the project should present all the appropriate conditions for the domain experts to properly estimate their activities, as mentioned above.

How to Estimate Duration

Accurate activity duration estimation is still the holy grail of project management. There are several methods on how to estimate tasks and they vary greatly based on the organization environment, the type of project, the type of activity, the resource pool available, historical data available and others. This section will highlight the important factors to keep in mind when performing activity duration estimation.

- The estimation should *always* be done by a domain expert and *never* by the project manager, unless the project manager is the domain expert *and* will be the resource implementing the activity
- Estimates should be submitted to a sanity check and be reviewed by independent domain experts
- Use analogous estimates (more on this later in this section) as much as possible
- Use three-point estimates (more on this later in this section) for activities that can be estimated using analogous estimate
- The project manager's job in estimation is:
  - o To provide the estimator with all needed information for the estimate to be generated without padding
  - o To capture and document all assumptions used in the estimation activity

o To formulate a reserve (more on this later in this section)

Analogous Estimate

As the name suggests, analogous estimate is done by comparison of the activity to be estimated with analogous activities that were executed in past projects. This can only be done if the organization has a system to capture project details that can be used as a reference during the activity duration estimation process. This type of estimation is the preferred one when available as it utilizes historical data as its basis, as opposed to expert judgment alone. This makes for an analytical data-driven estimate versus one that relies solely on the expertise of the estimator.

Three-Point Estimate

This is also called a PERT (program evaluation and review technique) estimate and is based on the fact that there is a very small probability of completing a given activity at the exact duration estimated. With this in mind, the estimator is basically asked to determine three estimates for the activity duration: a pessimistic one, an optimistic one, and a most likely one.

The pessimistic estimate considers the assumptions as being false and the identified risks becoming reality. This will obviously be the longest one of the three. The optimistic estimate assumes that everything will go exactly as planned and no roadblocks will be uncovered along the way. The most likely estimate will be what the domain expert would give if she were asked to flat out to estimate how long the activity would take, without being optimistic or pessimistic. This would be the so-called one-point estimate. Although these can be used in the construction of rough order of magnitude estimates, they should not be used when estimating the project activities duration.

With these three data points collected, the project manager will apply the following formula to derive the actual activity duration estimate that will be fed into the scheduling software application later on:

$$D = \frac{P + 4 \times M + O}{6}$$

D is the activity duration, P is the pessimistic estimate, M is the most likely estimate, and O is the optimistic estimate.

What this formula is saying is that a weighted average is being used and that the most likely estimate has a higher weight than the pessimistic and optimistic estimates. However, the formula does take those two estimates into account, providing an overall activity duration estimate that is more realistic than the one-point estimate.

As mentioned previously, it is statistically improbable that an activity will be completed within the exact estimated duration. For this reason, another important data point to be derived is the so-called activity standard deviation so the activity duration estimated range can be derived.

The activity range is the range of time that more realistically portrays how long it will take an activity to be completed; it is calculated as $D \pm SD$. The minimum time it would take to complete the activity would be $D - SD$ and the maximum time would be $D + SD$.

The standard deviation would be calculated by the formula $SD = \frac{P - O}{6}$, where P is the pessimistic estimate and O is the optimistic estimate for the activity.

Based on the information above, table 48 presents a suggestion for the information be summarized and captured as the output of the Estimate Time process.

ACTIVITY ESTIMATED TIME SUMMARY							
Activity Name	WP ID	Pessimistic (h)	Optimistic (h)	Most Likely (h)	Estimated Duration	Standard Deviation	Activity Range

Table 48. Activity List Example

The list shows all activities by name, the work package ID number the activity belongs to back in the WBS, the three-point estimate numbers in work hours, and the calculated duration, standard deviation, and activity range as presented by the previous formulas.

This information will be useful in determining the overall project schedule and the schedule range. Also, another important thing that should be looked for when analyzing the table above is risk identification. Activities that show a high standard deviation, meaning the pessimistic and optimistic estimates are too far apart, usually mean that either the estimator didn't have enough information to estimate the activity, thus the large standard deviation spread, or the activity carries risk that drove the estimator to widen the estimation range.

The project manager should use the table to flag those activities, and facilitate a team section that focuses on identifying the root cause of the wide estimate range to either lack of information or risks and properly handle it. If lack of information is the root cause, the project manager should revisit the activity and verify if further information can be obtained, allowing the activity estimator to take a second pass at the estimate. If the root cause is risk, the project manager should make sure those risks are included as part of the project risk register, as seen in previous sections.

The discussion above focused on time estimation; however, there is another equally important aspect of estimation, cost. The estimate cost process is where the cost of every activity in the project schedule is defined. The estimate cost process should include all project costs associated with the execution of the project. This process will generate information that will be used as inputs to the determine budget process, which aggregates the estimated cost of resources, WPs, and activities that will establish the cost baseline. There are mainly two methods available for cost estimation: analogous estimating and bottom-up estimating.

Analogous Estimating

The analogous cost estimating technique has the same basic idea as the analogous time estimate technique for activity duration as presented in a previous section. The main advantages of this estimating technique are the following:

- A less detailed WBS is needed for cost estimation. Since the project team is not executing a full-blown activity decomposition in order to derive the cost of every single activity and instead a cost is being derived by comparison with historical data on the WP level, the WBS doesn't have to include as much detail as needed for a bottom-up estimate.

- It is faster and less costly to create. Again, as there is no need to further decomposition of the WPs into their corresponding activity levels, the task of estimating become much simpler and less costly than the bottom-up counterpart.

On the flip side, there are some disadvantages with this technique:

- It is less accurate. Since the estimate is prepared with use of limited amount of detailed information, as the WBS is not fully flushed to the activity level, the overall estimate might be not as accurate should the WP prove to have some subtle differences in relation to the ones being used as baseline of comparison.
- It can't be applied to projects with uncertainty. This is somewhat obvious. If the project has a lot of uncertainties at the time the cost estimate process is executed, one can't really say whether the WPs can be really compared with historical ones or not.
- It does not take into account differences between projects. If all projects were exactly the same, then there wouldn't be any reason for the existence of the project manager profession, and the management of projects could have been turned into an automated task done by a software application. In reality, there are no two projects that are exactly alike; therefore, a comparison between WPs doesn't take that into account.

Bottom-Up Estimating

Bottom-up estimating is the technique where detailed estimating is done for each available project activity, as presented by the project schedule, and the estimates are rolled up into control accounts (more on this later in this section) and finally into an overall project cost estimate. This technique provides the following advantages:

- It is more accurate. Since the cost estimation is taking into account all the activities at their lowest level of granularity, the end result is a more accurate estimate.
- It gains easy buy-in from project team. This is for sure the easiest way to get buy-in from the project team on project cost as it participated in the tasks of defining the activities.
- It is analytical. Since the cost estimate comes from detailed analysis of the project, this option is certainly more analytical than the analogous estimating technique.

Some disadvantages of this technique are:

- It's time consuming and more expensive. Since all activities need to be very well detailed, this activity is much more time consuming and expensive than the analogous method of cost estimating.
- Analysis paralysis may occur. As mentioned, this method requires the project team to detail the activities to a very granular level. This might generate the famous analysis paralysis syndrome of project team members spending too much time to detail the activities. A project manager's ability to spot and prevent this from happening comes with time and experience.

Project Management Software

As mentioned in previous sections, there are many project management software applications that can be used to automate some of the most extraneous tasks, and activity costing is definitely one of them. Such software applications can be used so the activities are captured in a project schedule format with the corresponding estimated duration hours associated to them. The next step would be to enter the hourly rate for each one of the multiple resources. Then the software application automatically rolls that information into the multiple activity cost estimates.

Control Accounts

One may remember from the Create WBS section that a WBS is mainly a hierarchy of WPs that resemble an organizational chart. Figure 39 presents an example of WBS structure:

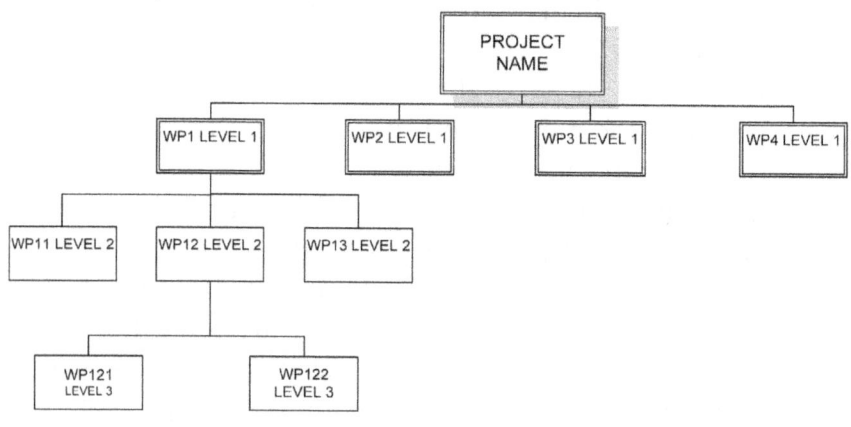

Figure 39. Control Accounts Hierarchy

It was mentioned during that section that the WBS contained multiple levels, and higher levels would include lower-level WPs associated with the same branch. As has been explored in the current section, the cost estimation activity will ultimately determine the cost of all WPs in the entire WBS.

Considering the case of large projects where thousands of WPs form the entire WBS, it might be a good idea to determine and flag some WPs that can be utilized as cost checkpoints for the activities around cost performance analysis that will be presented in later sections. These identified WPs can be at any level in the WBS and will show the aggregated cost that is basically the addition of the cost for all of their child WPs. These WPs receive the name control accounts.

There is no right or wrong way to determine the control accounts, and every project might have a different one based on its particular configuration.

The Project Schedule

Now that the sequence of activities and their dependencies have been established by the means of a network diagram and each activity has had its duration properly estimated, it is time to create the project schedule.

Each WP may have one or more activity, so the first step is to enter the activities and their estimated durations into their corresponding WPs. Once this task is completed, the project will have schedule time associated with the activities as well as an end date.

It is always good practice to take another look at the overall task dependencies once the activities have been plugged into the schedule. The first look was given at the WP level in order to create the network diagram; however, since the activities are more granular than the WPs, the project manager can look again for opportunities to shrink the schedule by potentially executing tasks in parallel if possible. This activity is called network analysis.

Critical Path

The critical path is the longest duration path through the network diagram that determines the shortest possible time to complete the overall project. Now that all activity durations are plugged into the network diagram, the critical path can be identified by the project manager. The critical path provides the project manager with important information that will help:
- Present the project duration to stakeholders
- Identify if any issues require immediate attention
- Identify the activities that can be delayed without delaying the project schedule
- Identify the project area that deserves the highest management focus

The method to finding the critical path is by identifying all possible flows through the network diagram and adding the activity durations along each path. The path with the longest duration is the critical path.

Along with the critical path, it is good practice to be aware of the duration for the alternative flow paths in the network diagram. When subtracting the critical path duration from the duration of each alternate path, the "float" is found. Float is basically the slack time that each of the network diagram flow paths has in relation to the critical path, or, in other words, how much each flow path can be delayed without delaying the project completion.

Once the float is calculated and understood, it can become a great project management asset. Once the float is know, the project manager can focus resources appropriately and better focus the management activities. For example, assume a project where the project manager knows the float is eighty hours and that a specific activity in the critical path is in risk of being delayed. A project resource that was allocated to execute an activity that was not part of the critical path can be real-located to help expedite the activity under risk on the critical path and prevent a project delay. The project manager knows that this maneuver can be done by up to eighty hours before the critical path changes.

Another important aspect of float is that it allows re-sources to be shared among multiple projects. The project team members, by knowing the dead line of their activities can better juggle multiple assignments and still deliver their work without creating project delays.

Moreover, the concept of float can be extended to the ac-tivity level. The float calculated for each network path will serve as input to the project manager to calculate how this float can be spread out among the activities that compose this path, based on their dependencies and resources.

There can be more than a single critical path on a project if more than one network path provides the same length in duration. This is usually not desirable as it does increase the project risk, as now the management focus needs to be split between multiple paths. Also, the critical path might change from one path to another. Two potential scenarios can be the causes of changes on the critical path. The first one is more ob-vious and happens when activities that are part of a noncritical path are delayed over their float time. This will cause this path to become the longest in the network diagram and therefore become the critical path. The second scenario is subtler and is the subject of the next topic.

Schedule Compression

It is important to keep in mind that even though the project schedule might be within the boundaries dictated by the project charter, the project manager should always be looking at ways to compress the overall project schedule and deliver project completion as early as feasible.

The definition of critical path suggests that any activity of schedule compression should be done by compression of the critical path. Therefore, the project team should creatively look at possible ways to either reduce the time the activities that compose the critical path can be completed or at ways to execute them in parallel, within the resource constraints of the project.

This activity can potentially move the critical path from one network path to another. This can happen in the case where schedule compression was successful in compressing the duration of the critical path in more than the float. This means that the path that was originally the critical path is no longer so and that the float is also changed. The new critical path and floats should be promptly identified and the process repeated with focus on the new critical path.

This process in intended to be iterative in nature and carried out by the project team and not by the project manager alone. This should be repeated until the team can't find alternatives to realistically compress the schedule any further. Once that point is reached, the project manager should cross-reference the project schedule against the schedule indicated at the project charter. If the project schedule is still not within the stakeholders' expectations captured at the project charter, the project manager should analyze alternate ways to bringing the project schedule to be within the one shown in the project charter. These alternate ways are:
  – Schedule Fast Tracking
  – Schedule Crash
  – Scope Reduction
  – Quality Reduction

Schedule Fast Tracking

This technique involves doing activities on the critical path in parallel that were originally planned in series once the initial schedule compression activities were completed. Fast tracking increases the risk of activity rework as some activities are started without having some of the inputs from activities that were originally scheduled to be fully completed prior to the beginning of those activities. For this reason, this technique might also add management time as it will require the project manager to pay extra attention to communication and flow of information between the team members involved in the fast-tracked activities.

Schedule Crash

This technique involves making cost and schedule trade-offs to achieve schedule compression for the least amount of incremental cost, that is, if the schedule must be compressed, what options will lead to the compression for the minimum additional cost? One obvious example is the insertion of additional resources to expedite certain critical path activities. This may be an acceptable solution if the project cost is still under budget as set by the project charter but the schedule is not within its acceptable boundaries.

Scope Reduction

Scope reduction is another way to compress the schedule. As you may remember, the project management triangle of constraints states that scope, schedule, and budget are interrelated and that changing one of the three vertices will induce a change on the other two. Reducing scope on strategic activities may achieve an overall schedule compression. However, scope reduction should be validated against the overall project charter and will definitely need approval from the change control board (the change control board will be explained in later sections). This should be a last resort by the project manager. If the scope can be reduced without an impact on the project charter, it means the project had scope creep prior to the scope reduction activity and was being poorly managed. What this means is that scope reduction to compress the schedule will always generate a change control event on well-managed projects.

Quality Reduction

Depending on the activity initial quality requirements, schedule compression can be potentially achieved by the reduction of these requirements. This is also a change that needs to be validated against the project charted and approved by the change control board. Much like the case of scope reduction, a quality reduction should be left as a last option to compress the schedule.

Reserves

It was shown in a previous section how each activity was to be estimated and how the standard deviation for the three-point activity estimation was to be taken into account for the activity ranges to be determined. So far, the current section has been referring to the activities as if they didn't have a range but were defined by a single duration value.

As it was seen, this cannot happen and will lead to problems if the activity ranges are not taken into account when determining the project schedule. Since each activity has a duration range, it is straightforward to understand that the project as a whole should also have a duration or uncertainty on the schedule.

Finding the overall project range is not as straightforward as finding the duration ranges for the individual activities. There is actually a process to be followed so the overall schedule range can be specified:

1) Find the expected duration for the project. This is done by adding all estimated activity durations for the activities that are part of the critical path.

2) Find the standard deviation for the project. Calculate the variances for each of the activities in the critical path. Variance is given by the following formula:

$$V = [\frac{P - O}{6}]^2$$

P is the pessimistic estimate and *O* is the optimistic estimate.

Add the individual variances for each activity in the critical path and take the square root of this sum. The final standard deviation for the project will thus be:

$$Psd = \sqrt{\sum V}$$

Psd is the total project standard deviation.

3) The overall project schedule range will thus *Pr* = *Pd* ± *Psd* where *Pr* is the project range, *Pd* is the calculated project duration, and *Psd* is the total project standard deviation.

The schedule needs to be always presented as a range to account for the stacked activity uncertainties. This will make up for a more realistic schedule that can be bought into by all stakeholders.

There are several project scheduling software packages available on the market that can be used to automate these calculations. The selection among them is a matter of preference by the organization and/or the project manager. One thing is extremely important to keep in mind, though: the scheduling software, much like any other project management software, is nothing more than a tool to be used by the project manager. Much like a hardware tool can be used as a weapon that kills or as an aid to an artist-carpenter who builds beautiful art pieces, the scheduling software tool is not what will create the schedule; the project manager is, and the software will only be as good as its utilization by the project manager.

Monitor and Control Schedule

One last topic that should be included as part of the Develop Project Schedule is the planning of how the schedule is to be controlled throughout the project implementation. The topic of controlling the schedule will be detailed in a future section; however, the high-level concept of it is that the schedule should be monitored constantly during the project implementation so the project manager has a chance to put plans in place in the event the schedule performance starts to slip. The project manager needs to apply a mechanism to measure the schedule performance against the planned baseline, identify if any variances are present, and, based on that, put plans in place to put the schedule back on track.

As it will be seen also in a future section, there is a technique called earned value analysis that provides a mechanism for the overall project performance to be measured against its planned baseline. There is a schedule component of this analysis, the so-called schedule performance index. For now, let's not get too bogged down on the definition of this index or how it is calculated, as it will be presented in detail later.

Table 49 provides a suggestion on how to capture this information as part of the schedule planning activities.

SCHEDULE CONTROL PARAMETERS				
Project Milestone	SPI Calculation Frequency	Response Plan Triggered If SPI $< X$	SPI Calculation Frequency After Response Plan	Response Plan

Table 49. Schedule Control Parameters

The table shows basically the frequency with which the SPI index is to be calculated during the implementation of each milestone, as well as what SPI results will eventually trigger a response plan. For example, assume that a project has ten milestones and that of those ten, four were identified as more involved and challenging than the other six. The project manager might want to monitor the schedule at a higher frequency rate during the execution of those four milestones than during the implementation of the less challenging other six.

The "Response Plan Triggered if SPI $< X$" basically shows the SPI result that will be the condition for the response plan being put in place. As will be explained in detail in a later section, an SPI value of one indicates a project that is on schedule; an SPI greater than one represents a project that is ahead of schedule; and an SPI value that is below one happens when a project is behind schedule. Utilizing the same example as before, the project manager might have a higher tolerance for the SPI number to slip below one during the implementation work of the more challenging milestone than for the SPI during the work to execute the other six milestones. In this case, the project manager might set the response plan triggered SPI value to be 0.90 for the challenging milestones and 1.0 for the less challenging ones.

The response plan column captures the actual response plan that will be implemented in case the monitored SPI for a given milestone slips to a lower value than the set trigger SPI value. Each milestone might have a different response plan, fast track, crash, resource leveling, and the others that were presented by this section, and that will be based on the type of work that is being executed for a given milestone.

The project manager might also elect to take another look at the other vertices of the project management triangle of constraints, scope, and budget, in case the SPI value of a given milestone gets below a critical threshold. For that reason, the project manager might elect to set multiple trigger SPI values for the same milestone that gave the implementation of different response plans based on different SPI values.

The last column that deserves a quick explanation is the one labeled "SPI Calculation Frequency After Response Plan." Once a response plan is implemented, the project manager might want to calculate the SPI index at a higher frequency rate than it has been calculated prior to the trigger condition being met. This is probably a good practice as it allows the project manager to know the level of effectiveness of the response plan sooner rather than later.

The Project Budget

One may rightfully ask, "Why is there a Develop Project Budget process since the project costs were already determined by the Estimate Cost process?" This is indeed a good question and its answer will come from the understanding of the difference between the project cost and the project budget.

The project cost, as defined in a previous section, is the cost of implementing all the WPs that comprise the project, including all costs of labor, equipment, and materials. The project budget is the total financial figure the organization will set aside to be used by the project and includes not only the project costs but also the reserves.

Reserves

In high-level terms, the project reserves are all moneys set aside to be used at the project manager's discretion to deal with risk responses. You may recall from the Perform Risk Response Planning section that the project risks were identified and qualified, and a cost estimate to respond to the risk was put in place. You may also remember that this cost estimate was basically defined as how much it would cost the organization to put in place the appropriate actions to respond to the risk accordingly. In other words, this cost estimate shows how much it would cost the organization if the risk actually became reality and needed to be handled by the project team.

There are two types of reserves: the contingency reserve and the management reserve. The contingency reserve is basically the reserve that will be allocated to deal with the risks as identified and qualified by the risk response plan. The project cost, as determined at the Estimate Cost process, added to the contingency reserve makes up for the project cost baseline and is the number the project manager will use as the cost to control the project against. The project cost performance calculations will use the cost baseline as its reference. Figure 40 illustrates this concept.

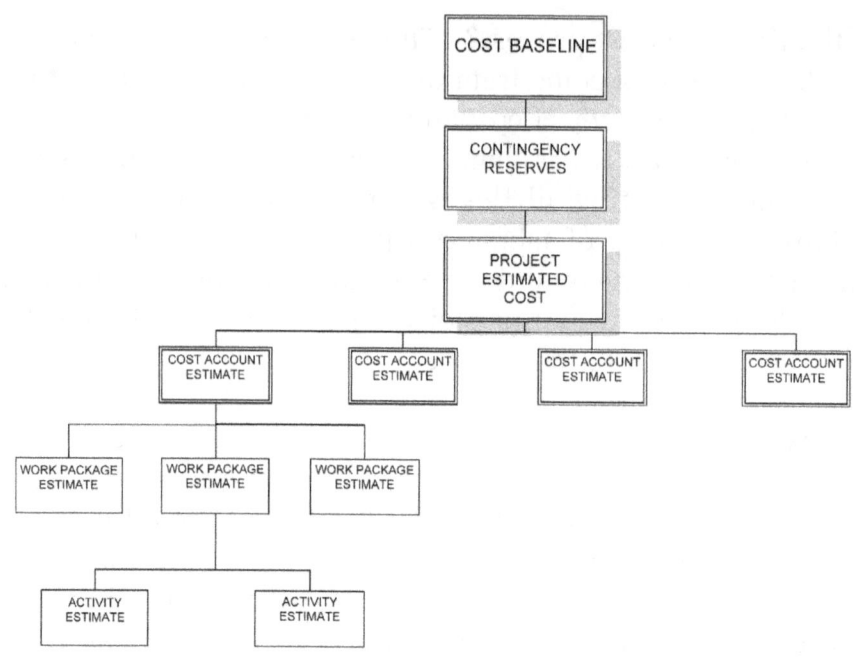

Figure 40. Project Cost Monitoring

The contingency reserve will then be calculated by adding together all the "Cost to Respond" values from the risk response planning table into a single lump sum.

It was mentioned in a previous section that a large chunk of the project manager's focus should be applied toward risk identification and mitigation. Therefore, well-managed projects will have a very comprehensive list of risks that the project manager and team will manage throughout the course of the project. However, every project has latent unforeseen risks, regardless of how well thought out the risk repose plan was by the project team. For instance, who could have predicted 9/11? That event for sure had a great impact, directly or indirectly, on live projects all around the world. What about the tsunami in Japan? Wouldn't one say that event caused impact on live projects also?

The management reserve is dedicated to those types of risks and their consequent responses. One may think that it would be very hard to determine the amount of money to be set aside as management reserve, and that is indeed the case. In fact, the amount allocated to management reserve will depend heavily on the type of project and past experience of the organization with these types of projects. Not only the project type will be a variable to this equation, but the industry type also plays a role in determining the reserve.

A common practice is to calculate the management reserve as a percentage of project cost baseline. In typical projects these percentages may vary from 3 to 10 percent for projects that can be somewhat analogous to historical projects done by the organization and from 5 to 25 percent if it is a totally new project. Another scenario that is important to consider is the T&M project that is supporting an NPI effort. In this case, the management reserve should probably be made even a little higher. There isn't really a guideline as to the percentage of project budget that should be allocated to the management reserve under these circumstances. It varies with several aspects, including but not limited to how stable the device under test is at the time the T&M project budget is being created, how technically challenging the T&M project is expected to be, how aggressive the overall NPI schedule needs to be, etc.

Thus, as illustrated by figure 41, the total project budget will be calculated as the cost baseline added to the management reserve.

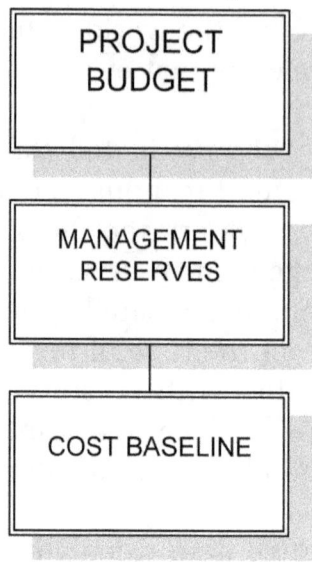

Figure 41. Project Budget Concept

Once the project budget is calculated, it is always a good project management practice to perform some sort of sanity check. Comparing this budget with the final similar project costs from the organization's project management system is an excellent validation task, if possible. Another good thing that can be done in that regard, in case the organization doesn't have a project management system or the project is new, is to run the numbers past other experienced project managers. Experienced project managers develop a knack for this type of task as they have lived through multiple projects over their careers and may have seen a project with some resemblance to the current one under analysis.

After all validation of the project budget is complete, there are two reconciliations that need to be done by the project manager: cash flow and project charter budget.

A cash flow analysis should be done as soon as the project budget is determined, and it entails understanding how the project costs will be spread over the duration of the project schedule. One simple example: if a given WP calls out the purchase of a piece of equipment by the end of month four of the project execution, the project manager needs to make sure the money will be made available by the organization for that purchase within that time frame. Involvement of upper management is definitely needed once the project manager has a clear picture on how the project will consume its budget.

Coming back to equipment purchasing example, if management states the money will only be available at the end of month five, the project manager needs to review and properly modify the project planning documents in order for that to be factored in. It is the project manager's job to provide a project baseline, prior to starting the project execution, which will contain realistic and achievable marks for scope, budget, and schedule. A project that is constantly being rebaselined is a project that most likely was poorly planned.

The reconciliation with the project charter is basically a cross-check to verify that the project budget determined by the planning activities is still within the constraints determined by the project charter. If the numbers don't jive, a meeting with the project sponsor and higher management is in order, at which the project manager will present the process behind determining the project budget as well as some alternatives around scope reduction, quality reduction, schedule expansion, or any other method the project manager identifies that can bring the project budget down within the project charter's constraints. Ultimately, it is management's decision to either grow the project budget on the project charter or select one of the presented alternatives to make sure the project budget aligns with the budget defined by the project charter.

However, it is the project manager's responsibility that this is properly presented to management's attention and that a realistic project budget is set at the end of the project planning processes. A project manager who is scared of this conversation with management and accepts the low budget dictated by the project charter is doing a disservice to the organization.

Monitor and Control Costs

One last topic that should be included as part of Develop Project Budget is the planning of how the project cost is to be controlled throughout the project implementation. The topic of controlling the cost will be detailed in a future section; however, the high-level concept of it is that the project cost should be monitored constantly during the project implementation so the project manager has a chance to put plans in place in the event the project cost performance starts to slip. The project manager needs to apply a mechanism to measure the cost performance against the planned baseline, identify if any variances are present, and, based on that, put plans in place in order to bring the schedule back on track. On top of those activities, the project manager should also be able to forecast the project cost at completion and compare it against the project budget set by the Develop Project Budget process.

As will be seen in a future section and mentioned in a previous section, the earned value analysis provides a mechanism for the overall project performance to be measured against its planned baseline. Similar to the case of the schedule, there is also a cost component of this analysis, the so-called cost performance index or CPI. As in the case of the project schedule, it is also important planning how this analysis will be performed, when and what will be the trigger mechanisms for response planning, and updating the estimated cost at project completion.

Table 50 provides a suggestion on how to capture this information as part of the cost planning activities.

COST CONTROL PARAMETERS						
Project Milestone	CPI Calculation Frequency	EAC Calculation Frequency	Response Plan Triggered If CPI $< X$	Response Plan Triggered If EAC $> \$$	CPI/EAC Calculation Frequency After Response Plan	Response Plan

Table 50. Cost Control Parameters Table

The table shows the frequency with which the CPI index and the EAC (estimate at completion) number are to be calculated during the implementation of each milestone, as well as what CPI and EAC results will eventually trigger a response plan. As in the schedule control parameters example presented, assume ten milestones. Of those ten, five were identified as having higher risk cost components than the other five. The project manager might want to monitor the cost and EAC at a higher frequency rate during the execution of those five milestones than during the implementation of the less risky other five.

The "Response Plan Triggered if CPI < X" and "Response Plan Triggered if EAC > \$" basically show the CPI index and EAC dollar amount that will be the conditions for the response plan to be put in place. As will be explained in further detail in a future section, a CPI value of one indicates a project that is on budget; a CPI greater than one represents a project that is below budget; and a CPI value that is below one occurs when a project is over budget. Utilizing the same example as before, the project manager might have a higher tolerance for the CPI number to slip below one during the implementation work of the riskier milestone than for the CPI calculated during the work to execute the other six milestones. In this case, the project manager might set the response plan triggered CPI value at 0.90 for the challenging milestones and at 1.0 for the less challenging ones. The same idea is applicable to the EAC amounts calculated.

The response plan column captures the actual response plan that will be implemented in case the trigger conditions are met. The project manager might also elect to take another look at the other vertices of the project management triangle of constraints, scope, and schedule, in case the CPI/EAC values of a given milestone surpass a certain critical threshold. For that reason, the project manager might elect to set multiple trigger CPI/EAC values for the same milestone that gave the implementation of different response plans based on those different values.

The column labeled "CPI/EAC Calculation Frequency After Response Plan" basically has the same function as its SPI counterpart column on the schedule control parameters table. Once a response plan is implemented, the project manager might want to calculate the CPI index and the EAC amount at a higher frequency rate than it has been calculated prior to the trigger condition being met for the same reasons the SPI might be calculated at a different rate once a schedule performance response plan has been put in place.

The next section will cover project risks, one of the most important aspects of any T&M project.

Risk Analysis

Risk analysis is an activity that, due to its importance, should be executed throughout the entire project life cycle. The first task that needs to be completed that will start off the risk analysis process is the definition of the risk register. The risk register is basically a summary table where all risks are not only identified at this early planning process, but also maintained within the entire scope of the project.

Table 51 presents the information that should be captured as part of the risk register.

RISK REGISTER							
Risk Name	Risk Status	Description	Category	Probability (1–10)	Severity (1–10)	Response Plan	Cost to Respond

Table 51. Risk Register Example

The first column is basically showing the risk name from which the risk will be identified and referred to. The risk name will come from the activity of identifying the risks. The risk status column will show an open, pending, or closed status. The open risks are the ones that haven't become active. The ones marked as pending are the ones that became active, its corresponding response plan was put in place, and now it is being monitored in order to verify that the response plan is indeed properly addressing the risk. The ones marked as closed are the ones that did become active, their corresponding response plans were put in place, and they were indeed addressed by their response planning, so the risk is no longer a threat.

Identify Risks

As it was mentioned previously, the method of identifying risks should include all stakeholders, as each stakeholder might look at the project from a different vantage point and be empowered by different backgrounds and interests. This diversity will be very beneficial to the activity and should be taken advantage of by the project manager, who should function as the facilitator. There are several techniques that can be applied in the identification of risks, such as:

- **Modeling.** Modeling, as seen previously, it is a great communication vehicle with which to engage the project stakeholders. Since modeling usually facilitates the analysis of exceptions and "what ifs," it is a natural driver in the identification of new risks that became apparent through the "what if" questions and answers.
- **Brainstorming.** This is usually done in meetings and it is advised that the common project team status meeting be made into a risk identification analysis brainstorming section as opposed to the usual going around the table status meetings. The risk brainstorming sections

are much higher value added for the expensive activity of having all members of the project team together in a meeting than a simple status update that can be obtained via much less expensive and rather more effective methods. The brainstorming section carried out as the project team meeting will keep the participants more engaged and will allow risks to be identified and analyzed by a panel composed of people from diverse backgrounds.

- **Interviews.** This consists basically of interviews done by the project manager or team with stakeholders and domain experts in order to potentially identify project risks.
- **Root-Cause Analysis.** Performing root-cause analysis of the higher-priority severity identified risks is always a good activity that helps the team to identify corollary risks that should also be captured as part of the risk register.
- **Delphi Technique.** This technique consists of sending requests for information to domain experts who can anonymously participate with their answers. Their responses are compiled and sent back to them for further review, until a consensus is achieved. This is a longer and impersonal process that is no longer used very often.

Description

This column shows a detailed description of the identified risk and should contain enough details so anyone who accesses the register, regardless of background and involvement with the project, can understand the high level of what the risk is about.

Category

A good practice to be implemented and included as part of the risk register is the risk category. There are multiple ways to categorize the identified risks; however, categorization by project area is probably the most useful. Under this method, risks can have one of the following categories: schedule, cost, quality, scope, resources, and stakeholder satisfaction. Having the register categorized this way allows the project manager to quickly filter the risks by project area while performing monitoring activities on each of the multiple project dimensions.

Probability and Impact

These two columns are self-explanatory and carry a number one for lowest probability/severity, a ten for the highest, and the spectrum between two and nine for an intermediate probability/severity scale.

Response Plan

The response plan should have a detailed explanation showing how the project manager plans to manage the project while taking the identified risk into account.

Cost to Respond

It is extremely important to capture the cost the project will incur should the response plan for a given risk be put in place. This number will be used by the project manager in compiling the management reserve numbers that will be added to the cost baseline to create the project budget.

The discussion above focused on qualitative risk analysis. The following will detail the quantitative analysis for risk.

It was mentioned above that risks ought to be prioritized into two categories; the first category will include all risks that are to be closely managed while the second group will show the risks that, though real, are to be monitored only. The rationale behind this is simple; imagine a large project whereby hundreds of risks were identified by the project team and other stakeholders. If the project manager were to address every single one of them, the project team wouldn't have time to actually implement the project deliverables as all they would be doing was going to be risk management.

Qualitative risk analysis is the process that allows the project team to properly sift through the identified risks and categorize them into the two risk types, creating a short list of risks that are worth the effort required to manage them closely.

Each identified risk is to be rated on probability of happening and severity to the project if it does happen. These ratings are to be organized in what is being called the probability and severity matrix, as presented in figure 42.

PROBABILITY	1	2	3	4	5	6	7	8	9	10
10										A
9	L									
8						O		F		
7		C					G		H	
6								I		
5			B							
4		K	J	N						
3					M					
2		D								
1	E									
	1	2	3	4	5	6	7	8	9	10
					SEVERITY					

Figure 42. Qualitative Risk Analysis Example

Each risk name is to be included in the corresponding matrix box that will match its probability and severity ratings, as shown above. Each letter on the matrix corresponds to a risk name captured in the risk register.

This matrix will give the project manager a visual representation of all captured risks. The next natural step is to determine what risks make which one of the two lists, the closely managed list or the monitor list. For that to happen, the risk matrix needs to be submitted to the qualification criteria, which is captured in the "Methodology" section of the risk management plan. Assuming the methodology section of the risk management plan states all risks above a rating of five for probability and severity will need to be closely managed, the matrix will immediately show the project manager what risks belong to which list, as displayed in figure 43.

PROBABILITY	1	2	3	4	5	6	7	8	9	10
10										A
9	L									
8						O		F		
7		C					G		H	
6								I		
5			B							
4		K	J	N						
3					M					
2		D								
1	E									

SEVERITY

Figure 43. Qualitative Risk Analysis Data

In this example, all risks that are within the red square belong to the closely managed list, and all risks that show within the yellow area belong to the monitor list. This information will aid the project manager in determining the risk response to all risks, information that is to be entered into the risk register. It is important to remember that even though the material is being presented in a sequential manner, project management is an iterative exercise where a later process might generate outputs that are to be used by previous processes as input.

Another important data point to be extracted from this matrix is the overall project risk. A project that shows the majority of the identified risks as being part of the red zone is a project whose viability and risk return ratio ought to be reevaluated. This analysis might show the project manager and project sponsor that the actual project risk might not be worth the investment and the project should be killed. This is actually a win for the organization as the project could have turned into a money pit if a good portion of those red risks had become reality and the project had been approved to move forward.

While the qualitative risk analysis is a subject evaluation, the quantitative analysis is an analytical data-driven process. As was mentioned previously, the risk register includes information on the cost of the risk response, and this cost will in turn serve as an input to calculate the management reserves for the project. The quantitative analysis is the technique used to calculate this cost.

Also, as mentioned, the risks identified and captured at the risk register are split into two categories, the ones that are to be closely managed and the ones that are to be monitored only. The ones to be closely managed are the ones that will contribute toward the management reserve and, therefore, should have their cost of response determined so that information is fed into the creation of the reserves. The risks that belong to the monitoring list don't need to have the quantitative analysis done to them.

Since we are dealing with probability events and not facts, determining the absolute cost of response would be not only difficult but would also have a good probability of over-shooting the need for management reserves, as the risk might or might not materialize. The alternative to determining the absolute cost of response would be to calculate the risk expected monetary value (EMV).

Calculating the expected monetary value is straightforward and is given by the multiplication of the probability rating by the cost of the impact on the risk on the project in case the risk materializes. In mathematical formalism, the expected monetary value is expressed as

$$EMV = P(\%) \times CosttoRespond(\$)$$

where P is the probability in percentage points. This number comes straight from the risk qualitative analysis process. A probability that received a rating of two corresponds to a 20 percent number plugged into the formula.

The cost to respond is not as straightforward to obtain as the probability. The cost to respond will heavily depend on the chosen risk response plan by the project team. Assuming an example where the response plan includes adding an extra resource to the project team and the risk materialization will add two hundred hours of work to the project, the cost of response would be calculated as the multiplication of the two hundred hours times the new resource hourly rate plus any recruiting time and other expense that are required to hire the new resource. Putting this example in numbers, assume the new resource hourly rate is $100 an hour and that the project manager will have to invest forty hours of her own time in recruiting activities plus another eight hours from multiple project team members in interviews. Assuming that the project manager's hourly rate is $135 per hour and that the project team members involved in the interview activities have a rate of $110 per hour, the total cost to respond will be given by:

$$CosttoRespond = 200 \, hours \times \frac{\$100}{h} + 40 \, hours \times \frac{\$135}{h} + 8 \, hours \times \frac{\$110}{h}$$

Assuming that the probability of this risk materializing is 20 percent, the expected monetary value will be calculated as:

$$EMV = 0.2 \times \$26{,}280 = \mathbf{\$5,256}$$

This amount is the one to be added to the EMV of the other risks that belong to the closely managed risk list and that will become the project management reserve. The example presented above is somewhat straightforward and easy to calculate. Other real-life examples might be a little more complex as the information of cost of responding might not be as readily calculated as the one presented. In such cases, the project manager and team should use assumptions and estimates in order to determine the cost to respond figure. The main idea that should be highlighted is that the quantitative analysis is what will give the project manager the cost that should be added so the management reserve is determined.

The definition of a risk in the project management world is a little different than the general definition of risk. In project management, a risk is defined as a probability event that might become a fact that can positively or negatively affect the project. The risks that can positively affect the project are called opportunities, and the ones that can negatively affect the project are called threats.

There are four main ways to respond to a threat risk:

- **Avoidance.** This method attempts to eliminate the threat by eliminating the cause, for example, relocate a WP to a different resource if the original resource is under risk of not being available at the time the WP will become active.

- **Mitigation.** This technique consists of attempting to reduce the probability or the severity of this threat, therefore making it a smaller risk that can be removed from the top risks to be managed and into the list of risks to be only monitored. One may remember that part of the methodology of managing risks is to multiply the probability and the severity column and determine a cutoff boundary number that would classify risks as risks to be closely managed and risks to be monitored. A reduction on either probability or severity might move a risk from the closely managed risk list onto the monitored risk list.

- **Transference.** This method basically attempts to make another party responsible for the risk. This can be achieved by purchasing insurance or warranties or by outsourcing the work. There is a typical link between the risk analysis activity and the make versus buy decision.

- **Acceptance.** There is still a fourth type of threat response, which is basically to do nothing and just accept the consequences of the threat if it does happen. Though one may be accepting the risk, contingency plans must be put in place to be implemented if the risk does become a fact, including the total cost of the response as well as added schedule time if applicable.

It is important to notice that one of the three techniques once applied might not eliminate the risk from the risk register altogether, but might change its probability-severity product or even morph the risk into a new one that needs to be properly captured and analyzed.

Like threats, opportunities also have methods to properly respond to them:

- **Exploitation.** Add extra work or change the project to some extent in order to maximize the chances of the opportunity to materialize.
- **Enhancement.** Take actions in order to increase the numbers of probability and/or severity.
- **Sharing.** Allocating ownership of the opportunity to a third party by the means of a partnership.
- **Acceptance.** The same concept as acceptance of a threat, presented above.

This information, along with the information determined by the qualitative risk analysis, will help the project manager and team to decide what is the appropriate response methodology for each of the identified risks.

It is important to highlight that though this discussion is being presented after the discussion on quantitative risk analysis, the response plan actually functions as input for the quantitative analysis to be performed. This shows the iterative nature of the risk management planning.

Figure 44 summarizes the activities to be performed during the risk management planning:

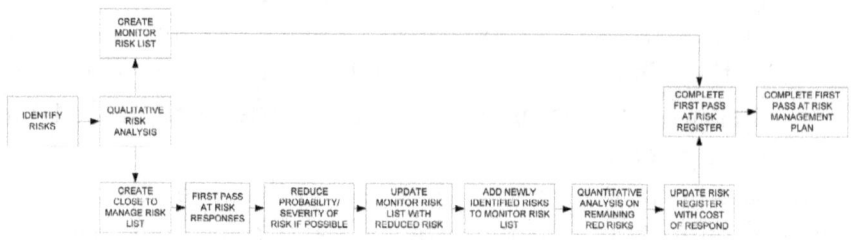

Figure 44. Project Management Planning Activities

Again, this process, although depicted as sequential, will have multiple iterations until the first pass at the risk register and management plan are completed.

Though a typical project life cycle doesn't stop once the processes detailed in this chapter have been completed, this is where this book will stop. The main goal of the book is to address the main root causes for T&M project failure, and the topics presented in the TMPM framework so far are the ones at the heart of the issues. This author understands that the activities to be performed to bring the project through completion, once what is being proposed by the TMPM framework is in place, can be executed in a typical manner. The T&M community, if supported by a framework such as the TMPM, will have a greater chance in succeeding on complex T&M projects.

CHAPTER 10:
Bringing It All Home

The introduction of the book started off the discussion on test and measurements projects. It showed how technological advances led to the increasing complexity of modern day-to-day end user products. The increasing complexity of our modern society drove test and measurements projects, the ones whose main mission is to build the systems that will assure the quality of such products, to an all-time high in complexity as well.

It introduced the reader to the main frameworks utilized in the execution of such projects: the PMBOK® by the Project Management Institute, the systems engineering (SE) framework by INCOSE, and the agile methodology. The stage was set for an analysis of each of these frameworks in a later chapter, in an attempt to determine whether the problems encountered in T&M projects were due simply do a lack of training of project managers and project team members on such frameworks, or if the problem had deeper roots than that.

The main goal for section 1 was to properly frame the problems around test and measurements projects and come up with, if possible, an addressable problem statement to be tackled.

It started by involving the reader in a discussion on the main reasons why test and measurements projects fail. A root-cause analysis of data collected from hundreds of test and measurements projects over a span of fifteen years was presented. This analysis culminated with the two main root causes for test and measurements project failure: lack of well-established requirements and poor project planning. The first step for addressing these issues was to apply the five-whys technique and drill deeper into their core. Underlying issues were presented, those being the main drivers of the two root-cause conditions for project failure. The T&M problem statement then stated the need to address these underlying issues for the two project failure root causes.

The next chapter focused on engagement of system integration companies in the execution of T&M projects. Since there are literally thousands of such companies offering services in the T&M industry, the root-cause analysis, in order to be thorough and complete, needed to include the relationship between clients and these companies. Five main issues in dealing with integrator companies were presented, which were then added to the overall T&M problem statement. To increase the odds for the success of these projects, organizations needed to address the two main drivers of T&M project failure, as well as address these five main issues when engaged with a system integration firm.

The section went on to present a detailed account of each of the two main T&M execution frameworks, the PMBOK® and SE frameworks. The goal for this chapter was to set the stage for a gap analysis of the frameworks against the root-cause underlying issues introduced in a previous chapter. This gap analysis determined that neither framework nor the agile methodology could stand alone and be fully applied, without modifications, to address the underlying issues in question.

The last chapter of the section wrapped up the analysis of the frameworks by cross-referencing them against the issues with engaging system integration companies. As could be expected, the proper application of the frameworks is not the answer to a successful collaboration between client and integrator.

The conclusion to be drawn from this analysis is that there isn't a framework in the industry today that can fully address the idiosyncrasies of T&M projects. Although the requirements and complexity of such projects have evolved, the way professionals go about executing them still has historical ties to frameworks that no longer address the needs of these projects. This section, however, identified the strengths of each framework, which suggested that they do have merit and probably just need to be tailored to be best applied in T&M projects.

This is the whole focus of section 2 of the book. The TMPM framework was presented in detail. The TMPM framework is a combining of the PMBOK®, SE, and agile methodology into a comprehensive framework, tailored for T&M projects.

The first chapter of this section started off the discussion by presenting a typical landscape of test and measurements projects when it comes to project team and project stakeholders. It showed how diverse the technical backgrounds of the project team need to be to successfully tackle the modern complexity of such projects. It also detailed how, due to the important and pivotal role a test system plays in any organizational structure, the project manager needs to interface with several stakeholders with different skills and departments, from the test operator all the way to C-level stakeholders. This discussion came to the conclusion that the two root causes of project failure were actually driven by human aspects: poor communication with stakeholders and missing stakeholders. These are actually the two problems that needed fixing. It showed how, by addressing the stakeholders issues, the two root causes presented in section 1 were actually being addressed.

This chapter set the stage for the presentation of a modified organizational structure for the execution of T&M projects. Under this structure, the responsibility to manage the project is split into two roles, the systems engineer and the project manager. The systems engineer would be the one connecting with the technical stakeholder and providing technical leadership to the project team. The project manager would be the one with a personality that would allow her to connect, on a human level, with the multiple stakeholders that are typical to T&M projects. It was shown how this structure provides the organization with a much better chance in addressing the human resource issues that are usually present when the old paradigm of a single technical project manager is applied.

This chapter also presented a new role, a liaison between client and integrator. This person would bridge the gaps between the two organizations and make sure the business value of the test system being built is maximized for the client. It showed how this person can contribute to address the five main issues that can be present in the engagement between client and integrator.

The next chapter focused on the tools utilized by the framework. It presented an introduction to UML and showed how to best use UML to execute system modeling. It showed how system modeling, at multiple levels of the project life cycle, helps the project management body address the stakeholder issues that were identified as the true root causes of T&M project failure.

The last chapter of this section brought it all together into a set of tasks, or processes, that would utilize the organizational structure and the tools presented by the previous two chapters.

It is this author's sincere hope that what was presented in this book contributes somehow to the overall improvement of T&M project results. Even though these projects may not be as glamorous as product development projects, which are directly connected to the improvement of modern life, their failure can easily drive product development initiatives to the ground. Therefore, they do have a direct impact on the improvement of everyday life.

About the Author

Filipe Altoe is a natural of Brazil, where he obtained his BS in electrical engineering and MSc in control systems. Filipe started his career as a controls engineer for a pharmaceutical company, when he had his first exposure to project management of technical projects as well as failed projects.

Filipe moved to the United States in 2000 to work for National Instruments. He left NI and kept up his involvement in the test and measurements industry. Having worked as a director of product development in charge of the organization's new product introduction, and as an executive in a services organization in charge of implementing test systems for clients, Filipe has been able to collect and compile data on dozens of complex test and measurements projects from his seventeen years in the industry.

Filipe is an active member of the project management community, and he is known as a thought leader in the arena of project management for complex technical projects. Filipe has been featured in *PM Network* magazine, the flagship worldwide monthly publication by the Project Management Institute (PMI), the world's leading professional association for project management. He also had the opportunity to be the keynote speaker for several PMI chapters. Filipe holds the professional project management (PMP) certification by PMI and certification on National Instrument technology by NI.

Filipe currently runs his own test and measurements company (www.TSXperts.com), and he focuses on helping clients to successfully implement their test projects by offering services of project rescue, support to new product introductions, and full system integration of test projects. Filipe can be reached at filipe@tsxperts.com.

References

[1] PMI Project Management Body of Knowledge (PMBOK®), Fourth Edition.

[2] INCOSE Systems Engineering Handbook, version 3.

[3] Analysis of DOD 5000.2R Project Management Process, Robin E. Whitworth, December 1998.

[4] Managing Complex Technical Projects: A Systems Engineering Approach, R. Ian Faulconbridge and Michael J. Ryan.

[5] PDMA Handbook of New Product Development, Second Edition, Kenneth B. Kahn.

[6] UML Distilled: A Brief Guide to the Standard Object Modeling Language, Second Edition, Martin Fowler and Kendall Scott.

[7] UML for Systems Engineering: Watching the Wheels, Second Edition, John Holt.

[8] Determining Project Requirements, Hans Jonasson.

[9] IIBA Business Analysis Body of Knowledge, version 2.0, International Institute of Business Analysis.

[10] Agile Project Management: How to Succeed in the Face of Changing Project Requirements, Gary Chin.